This book is to be returned on
or before the date stamped below

1 8 APR 2002

2 3 MAY 2002

1 3 NOV 2002

1 4 FEB 2003

1 0 MAR 2003

- 9 JUN 2003

Conservation and the Use of Wildlife Resources

CONSERVATION BIOLOGY SERIES

Series Editors

Dr F.B. Goldsmith
Ecology and Conservation Unit, Department of Biology, University College London, Gower Street, London WC1E 6BT, UK

Dr E. Duffey OBE
Cergne House, Church Street, Wadenhoe, Peterborough PE8 5ST, UK

The aim of this Series is to provide major summaries of important topics in conservation. The books have the following features:

- original material
- readable and attractive format
- authoritative, comprehensive, thorough and well-referenced
- based on ecological science
- designed for specialists, students and naturalists

In the last 20 years **conservation** has been recognized as one of the most important of all human goals and activities. Since the United Nations Conference on Environment and Development in Rio in June 1992, **biodiversity** has been recognized as a major topic within nature conservation, and each participating country is to prepare its biodiversity strategy. Those scientists preparing these strategies recognize **monitoring** as an essential part of any such strategy. Chapman & Hall has been prominent in publishing key works on monitoring and biodiversity, and aims with this new Series to cover subjects such as conservation management, conservation issues, evaluation of wildlife and biodiversity.

The Series contains texts that are scientific and authoritative and present the reader with precise, reliable and succinct information. Each volume is scientifically based, fully referenced and attractively illustrated. They are readable and appealing to both advanced students and active members of conservation organizations.

Further books for the Series are currently being commissioned and those who wish to contribute to the Series, or would like to know more about it, are invited to contact one of the Editors or Chapman & Hall.

Conservation and the Use of Wildlife Resources

Edited by
M. Bolton

CHAPMAN & HALL
London · Weinheim · New York · Tokyo · Melbourne · Madras

Published by Chapman & Hall, 2–6 Boundary Row, London SE1 8HN, UK

Chapman & Hall, 2–6 Boundary Row, London SE1 8HN, UK

Chapman & Hall GmbH, Pappelallee 3, 69469 Weinheim, Germany

Chapman & Hall USA, 115 Fifth Avenue, New York NY, 10003, USA

Chapman & Hall Japan, ITP-Japan, Kyowa Building, 3F, 2-2-1, Hirakawacho, Chiyoda-ku, Tokyo 102, Japan

Chapman & Hall Australia, 102 Dodds Street, South Melbourne, Victoria 3205, Australia

Chapman & Hall India, R. Seshadri, 32 Second Main Road, CIT East, Madras 600 035, India

First edition 1997

© 1997 Chapman & Hall

Typeset in 10/12pt Sabon by Acorn Bookwork, Salisbury, Wilts

Printed in Great Britain at the University Press, Cambridge
ISBN   0 412 71350 0

A catalogue record for this book is available from the British Library
Library of Congress Catalog Card Number: 96–72148

∞ Printed on permanent acid-free text paper, manufactured in accordance with ANSI/NISO Z39.48-1992 and ANSI/NISO Z39.48-1984 (Permanence of Paper).

Contents

Contributors

Nicholas J. Aebischer
The Game Conservancy Trust,
Fordingbridge,
Hampshire SP6 1EF, UK

Melvin Bolton
PO Box 879,
Yeppoon,
Queensland 4703, Australia

Owen Lee Griffiths
Bioculture (Mauritius) Ltd,
Senneville,
Rivière des Anguilles, Mauritius

John S. Lucas
Zoology Department,
James Cook University,
Queensland 4811, Australia

Tim R. New
School of Zoology,
La Trobe University,
Bundoora,
Victoria 3083, Australia

Pete Reynolds
Motacilla,
2 West Point, Garvald,
East Lothian EH41 4LN,
Scotland, UK

Brian Staines
Institute of Terrestrial Ecology,
Hill of Brathens, Glassel, Banchory,
Kincardineshire AB31 4BY,
Scotland, UK

Mary-Ann Stanley
Bioculture (Mauritius) Ltd,
Senneville,
Rivière des Anguilles, Mauritius

Michael H. Woodford,
IUCN/SSC Veterinary Specialist Group,
2440 Virginia Avenue, NW,
Apt. D-1105, Washington, DC 20037, USA

Preface

Not everybody cares about the fate of wild animals or the state of the natural environment. I met a lady who said it wouldn't worry her if all the wild animals in the world disappeared overnight. She was a city person, she said. There are also people who would prefer to let animals become extinct than to have them kept in captivity – no matter how progressive the zoo. There are those who, on principle, will not eat meat, let alone do the killing, and there are those who enjoy nothing so much as shooting birds. People in the last two camps may oppose each other in claiming to be conservationists.

Extremists are unlikely to find their opinions being reversed by this book but, because of the scope of the subject, I believe there is a good chance that anybody with an interest in wildlife will find in it something new to think about. It may not be too much to hope that a few disagreements might also be settled because I suspect there is more common ground than is generally realized among those with opposing views.

The term 'wildlife' lacks precision but when we need to refer collectively, and loosely, to wild animals living in a wild state it seems as good a word as any other. The buzzword 'biodiversity' also lacks precision, but in its fullest expression it takes in the genetic make-up of all the different assemblages of the millions of species that constitute life on earth. Wild animals in their natural habitats are, of course, selections of biological diversity.

The justification for trying to conserve biodiversity rests on ethical, aesthetic and purely utilitarian grounds. The rationale has been explored and summarized in many publications (e.g. WCMC 1992) and no space will be given up to it here.

Although people have always made use of wild animals, the scale and complexity of usage are far greater now than in primitive times. There are also broader social divisions between those who use wildlife directly and those who do not; between urban and rural societies; between rich and poor nations. These divisions are partly responsible for the sometimes conflicting opinions about what should be done under the banner of conservation.

The conservation value of using wildlife sustainably must take into account the extent to which usage improves the chances of protecting land from being converted for other purposes. It will be argued that all options must be kept open and that in some circumstances conservation through utilization will be an appropriate alternative when total protection cannot be achieved. If there was a realistic option of maintaining all our biodiversity by leaving wildlife undisturbed, there would be no reason for this book.

Throughout the book we try to maintain a global perspective as we examine the relationship between using wildlife and conserving it. At the global level, it is possible to make some generalizations. It is true, for example, that the main threat to wildlife everywhere is the loss of habitat. Wilderness is everywhere under pressure, and wildlife is being displaced out of existence rather than being hunted to extinction. But locally, or with regard to particular species, the circumstances can be very different. Some species are more threatened by selective hunting than by habitat loss; some species are increasing in number; others are mainly threatened by aliens – animals from another continent. And the animals themselves, in all their forms and with all their lifestyles, vary enormously in their capacity to cope with the various kinds of threat. Sound generalizations, therefore, are harder to find than might be thought.

The World Conservation Strategy (IUCN/UNEP/WWF 1980) stressed the importance of sustainability in the use of renewable resources and this was echoed in the follow-up document, *Caring for the Earth* (IUCN/UNEP/WWF 1991). However, neither publication went so far as to *promote* the use of wildlife as a means of conserving it. For a long time it was generally accepted that encouraging the use of wildlife was likely to speed up its destruction and could not possibly be adopted as a conservation strategy – especially by organizations dependent on public donations. But the difficulties of protecting wildlife and wild places from human activities continue to increase. Demands for wildlife to pay its way have become more insistent. The 'use it or lose it' philosophy has gained strength. In 1990, at the 18th General Assembly of the World Conservation Union (IUCN) held in Perth, Australia, it was resolved (Recommendation 18/24) that '...ethical, wise and sustainable use of some wildlife ... can be consistent with and encourage conservation, where such use is in accordance with adequate safeguards...' (IUCN 1991).

The resolution was understandably cautious and carefully worded. It was recommended that guidelines should be developed on the sustainable use of wildlife. Draft guidelines have since been produced but have not been adopted by IUCN (Prescott-Allen and Prescott-Allen 1996). Not only is the issue still a contentious one but the pros and cons of conserving wildlife by using it are almost specific to every case. Like many complex issues, however, the controversy has tended to become polarized, as though making use of wild animals were altogether a good thing or a bad thing for conservation. The purpose of this book is to show that, in reality, it all depends....

In Part One the broad scope of the subject is explored. Questions of purpose, scale of use and sustainability are considered in relation to ecological, socio-cultural and economic circumstances. The concept of ecological sustainability is given some emphasis but not to the exclusion of other aspects of sustainable use.

Part Two, with contributions from specialist authors, takes a closer look at some particular categories of use and includes a selection of case studies which involve very different taxa. In Chapter 5 John Lucas describes attempts to bring sustainability to the exploitation of giant clams in the South Pacific. Tim New, in Chapter 6, shows how an international trade in butterflies can benefit both the butterfly species and the people of the forests where the butterflies live. In Chapter 7 I review progress with crocodile ranching as a strategy to restore depleted crocodile populations in tropical waters and wetlands. Nicholas Aebischer, in Chapter 8, analyses the conservation benefits of maintaining the grey partridge as a game bird in the English countryside. My intention in Chapter 9 is to show how expertise, developed through the use of wildlife, has contributed to conservation – in this case through falconry. In Chapter 10 Pete Reynolds and Brian Staines take a critical look at the way in which red deer herds are managed in the Scottish Highlands. Owen Griffiths and Mary-Ann Stanley explain in Chapter 11 how the exploitation of crab-eating macaques, an introduced species which has become a pest in Mauritius, can be turned to conservation advantage. Michael Woodford, in Chapter 12, reveals the veterinary implications of using wildlife, especially as food for people. I take a look at the links between conservation and captivity in Chapter 13, and in Chapter 14 I offer an overview of ecotourism as a non-consumptive, but potentially high-impact, use of wildlife.

Part Three consists of a single chapter in which there is an attempt to synthesize; to make such generalizations as can be supported; and to identify the criteria for success in using wildlife for the benefit of conservation. In doing this I draw heavily on the chapters in Part Two. The authors of that section may not agree with everything I say, and only I can be responsible for any errors I may have made. This book would not have been possible without the contributing authors, all of whom supplied chapters in good faith. I can only hope that no author feels the trust was misplaced.

References

IUCN (1991) *Proceedings of the 18th Session of the General Assembly of IUCN.* Gland, Switzerland.

IUCN/UNEP/WWF (1980) *World Conservation Strategy: Living Resource Conservation for Sustainable Development.* Gland, Switzerland.

IUCN/UNEP/WWF (1991) *Caring for the Earth. A Strategy for Sustainable Living.* Gland, Switzerland.

Prescott-Allen, R. and Prescott-Allen, C. (1996) *Assessing the Sustainability of Uses of Wild Species – Case Studies and Initial Assessment Procedure.* IUCN, Gland, Switzerland, and Cambridge, UK.

WCMC (1992) *Global Biodiversity: Status of the Earth's Living Resources.* Chapman & Hall, London.

Acknowledgements

Anyone writing a book of this sort, while living in comparative isolation, needs a lot of help. I was particularly fortunate from the outset in having the solid support and encouragement of the Series Editor, Barrie Goldsmith. I could depend, as well, on Bob Carling and Helen Sharples in Chapman & Hall's Life Sciences office. Jacqui Morris kindly provided valuable contacts at the outset and Paul de Fritaes generously provided a forwarding service for the numerous publications which had to be obtained from the UK. Throughout the book it will be my pleasure to acknowledge the help I have received with respect to particular chapters.

Part One

Perspectives and General Issues

1

Subsistence use of wildlife

1.1 HUNTING AND GATHERING: THE ORIGINAL LIFESTYLE

Our pre-human ancestors presumably hunted as well as gathered. Chimpanzees (*Pan troglodytes*) from whose ancestors the human line diverged some 8 million years ago, are known to hunt cooperatively for red colobus monkeys (*Colobus badius tephrosceles*) and the procedure has been the subject of a remarkable film by the BBC Natural History Unit. Exactly when human beings became effective hunters of big game, as well as catchers of small creatures, is a separate question but game was to become especially important for man because early humans (probably including the Neanderthals) were able to make use of the skins for clothing.

As humanity approached its agricultural revolution about 10 000 years ago, there were between 5 and 10 million human beings on earth (Keyfitz 1989) and they are known to have consumed a wide variety of animals and plants from their hunting and gathering. By that time numerous uses had been discovered for other non-edible parts of animals as well as skins. Bones, horns and teeth were used for making tools, weapons, utensils and carvings (ivory). Possibly, even before agriculture, people had discovered the value of having tame dogs along for the hunt in what was the first use of live animals by man (ants had been farming aphids for a very long time).

Human fertility and mortality rates must both have been high because people's numbers were controlled by the same ecological opportunities and constraints that regulated the population densities of all large, omnivorous mammals. In that respect, *Homo sapiens* was little different from the boars, bears or baboons which foraged under similar circumstances. This may seem like a statement of the obvious but these basic facts are fundamental to the subject of present-day conservation; some serious misunderstandings persist largely because these simple facts are overlooked.

Conservation and the Use of Wildlife Resources. Edited by M. Bolton.
Published in 1997 by Chapman & Hall. ISBN 0 412 71350 0.

1.2   RECENT AND EXISTING HUNTER–GATHERERS

A few ethnic groups persisted as hunter–gatherers, mainly in areas not readily suitable for agriculture. However, today there can be very few people who have no contact with modern civilization, though their involvement may be limited to the use of a few metal objects such as tin cans and steel axes. The situation was succinctly described by Harcourt and Stewart (1980) with reference to the hunter–gatherers of Gabon: 'They have their radios, stainless steel cutlery and flowery painted crockery, and yet, if you arrive in one of the small forest villages before about 10 a.m., no-one will be there: the men are out hunting and the women gathering the fruits of the forest.'

The diets and survival strategies of the more recent and existing hunter–gatherers are well known compared with those of prehistoric man. None has been exclusively vegetarian, although opportunities for eating red meat may have been irregular; Bushmen of the Kalahari sometimes go without meat for weeks at a time (Silberbauer 1972). At the other extreme, the Inuit groups of the Arctic ate the meat of marine mammals, together with fresh-water-stream fish, as their staple. In general, hunter–gatherers are opportunists, making do with whatever is available during hard times but taking full advantage of times of plenty – when they are also able to show their preferences by being highly selective.

1.2.1   Australian Aborigines

Until Europeans came to Australia, 200 years ago, the indigenous people were exclusively hunter–gatherers and, despite tribal and other social factors which influence human distribution patterns, it can be assumed that they lived in densities which reflected the carrying capacities of the different ecological zones (section 1.3). Although these people practised no cultivation, they used fire extensively and deliberately in order to improve productivity by stimulating vegetation growth (Kirk 1981).

There is no complete analysis of the diet that Aborigines obtained entirely by traditional methods and the opportunity for making such records has now gone. Nevertheless, from studies of groups making minimal use of non-traditional supplies, it has been found that the intake of animal protein varies markedly between regions and seasons. In arid areas 70–80% of the bulk of food is likely to be vegetable, but at a late dry season camp in Arnhem Land, vegetation accounted for less than 10%, fishing provided 20% and the remaining 70% came from hunted kangaroos (Kirk 1981). A study by Meehan (1977) among Anbarra people in coastal Arnhem Land showed a very comprehensive use of the environment (Figure 1.1). A similarly varied diet has been recorded for bushman groups of southern Africa (Borshay 1972; Silberbauer 1972).

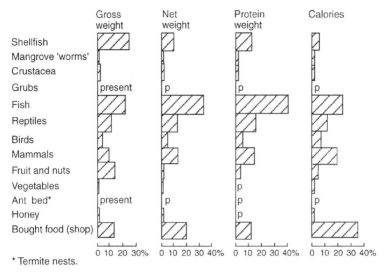

Figure 1.1 Diet of the Anbarra Aborigines during a one-month period. (From Meehan 1977.)

1.2.2 Hunting and gathering in tropical forests

Wildlife from tropical forests still feeds millions of people. Communities living near forest in Nigeria were reported to obtain 84% of their animal protein from bushmeat and in Ghana about 75% of the population regularly eats meat from wild animals (Asibey and Child 1990). Small animals, including most birds, rodents, snakes, lizards, amphibians, snails and insects, make up the protein bulk in subsistence diets but larger mammals, such as forest antelopes and primates, are likely to be sold by the hunter (section 1.2.4). The people of these regions are settled agriculturalists, no longer living as hunter–gatherers, and yet wildlife as food is still extremely important to them.

Numerous authors have recorded the range of species used for food in South America and have calculated the contribution that wildlife makes to the diets of different ethnic groups. Redford and Robinson (1991) list nearly 30 species of mammals and birds which are commonly taken. In summary these authors point out that hunters generally take more mammals than birds and more birds than reptiles. This conclusion is supported by Mittermeier (1991) who lists the mammals, birds and reptiles (21, 7 and 4 genera respectively) taken by four hunting groups in Suriname.

In the African territory of Rio Muni, in Equatorial Guinea, 30 species of mammals, including the endangered primates, gorilla (*Gorilla gorilla*) and drill (*Mandrillus leucophaeus*), were identified in the main markets (Fa *et al.* 1995).

1.2.3   Hunting selectivity

The discrepancy between kill rates for preferred species and the species' abundance often shows pronounced selectivity in hunting. In one Indian territory in Equador, woolly monkeys (*Lagothrix lagothricha*) were killed six times more frequently than agoutis (*Dasyprocta* sp) over a 10-year period (Vickers 1991). It is extremely unlikely that the monkeys would be six times more common. Whatever other factors may be involved (e.g. ease of location and dispatch), the 8-kg primate obviously represents a better return for a shotgun cartridge than does a 2-kg rodent. Over the 10-year period, agouti kill rates tended to be highest when overall hunting yields were low, further suggesting the use of the rodent as a makeweight species. Among Neotropical primates, woolly monkeys appear to be particularly vulnerable to subsistence hunting as they are targeted to provide a 'wild-meat bonanza' when new forest sites are opened up, but soon become locally extinct (Peres 1991).

In the Bolivian Amazon, a settled community of Yuquí Indians was forced to resort to smaller, less-preferred species as wildlife became depleted in the vicinity of the settlement, and neighbouring forest was converted to agriculture (Stearman and Redford 1995). The trend was conspicuous over the 5-year period from 1983 to 1988 during which daily protein consumption dropped from 88 g to 40 g per caput. In 1983 the intake comprised 156 animals from 27 taxa but in 1988 the animals numbered 348, and 44 taxa were being exploited.

On the north coast of Kenya, subsistence hunters were found to be harvesting wild mammals with a total biomass of 350 kg/km² from a forest of 372 km². In the past the hunters killed large mammals, including elephants (*Loxodonta africana*), black rhinos (*Diceros bicornis*), buffalo (*Syncerus caffer*) and bushbuck (*Tragelaphus scriptus*). They no longer kill elephants for meat because of the high risk of being caught and punished under the law. Rhino have already disappeared and the remaining large mammals have been harvested at such high levels that they are now rare in the forest. Instead, two species of elephant shrew (*Rhynchocyon chrysopygus* and *Petrodomus tetradactylus*) comprise 65% of all kills – although they make up only a small percentage of the total edible biomass. The bulk of the meat is obtained from relatively small numbers of bushpigs (*Potamochoerus porcus*) and aardvarks (*Orycteropus afer*). Most of the meat is for home consumption, rather than the market. Hunting with dogs and bows occurs throughout the forest but the shrews are caught in hundreds of traps which are set daily; mostly within 2 km of the forest edge (Fitzgibbon *et al.* 1995). A similar combination of shooting and snaring was described in west Cameroon, where dogs are used to drive primates, such as the endangered drill (*Mandrillus leucophaeus*), into trees, from where they are easily brought down with shotguns (King 1994).

The hunters in these examples appear to be killing preferred animals for as long as possible and resorting to less preferred species as they are forced to do so. Such evidence as is available supports the view that subsistence hunters try to satisfy each day's needs in accordance with foraging theory, rather than practising selective restraint for long-term conservation of the preferred species (Alvard 1993).

In the Equatorial Guinea study cited above, three species comprised nearly 70% of all carcasses in markets serving the administrative capital, Malabo. These were the blue duiker (*Cephalophus monticola*), Emin's rat (*Cricetomys emini*) and the brush-tailed porcupine (*Atherurus africanus*). The frequency of species in the markets did not correspond to their observed density in the forest and it was assumed that marketing considerations influenced the selection of hunters' prey.

First estimates of sustainable harvest rates for marketed animals were worked out following a method developed by Robinson and Redford (1991). The method necessarily involves some assumptions. In particular, it is assumed that population growth of forest mammals is density dependent (Chapter 3), and that the sustainable offtake of forest mammals is related to their average lifespan. The method has proved useful in attempts to compare the impact of hunting on different forest mammals. It was never intended that the method should be used to generate single-species harvesting schedules. Figure 1.2 shows the extent to which actual (marketed) harvests deviate from the calculated sustainable harvests for animals commonly appearing in the meat markets of the island of Bioko (formerly Fernando Po) in Equatorial Guinea.

1.2.4   Impact of marketing

As Hames (1991) had declared: 'entrance into a market economy is undoubtedly the most potentially devastating change in aboriginal environmental relations'.

The numerical relationship between hunters and wildlife living in a given area becomes irrelevant if the hunters are serving an extended area through markets. In west Cameroon, traders regularly attend village markets to buy bushmeat for restaurants up to 150 km away (King 1994). The market, in effect, opens a one-way channel through which wildlife is converted to money. It is a one-way channel because subsistence hunters usually have no way of converting money to wildlife except through increased offtake.

In Equatorial Guinea it was noted that meat extraction from the bush was proportional to the purchasing power of the urban markets. The volume of bushmeat serving Malabo, the administrative capital in Bioko, was 70% greater than that being sold through Bata, in Rio Muni, though there was little difference in population size between the two centres (52 000 and 55 000 respectively) (Fa *et al.* 1995).

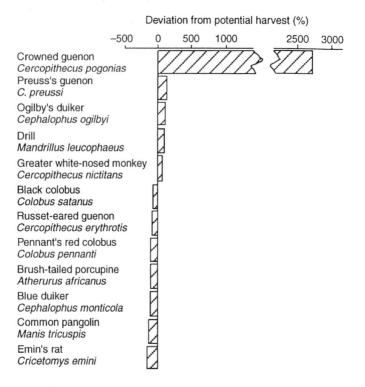

Figure 1.2   Deviation of actual harvests from calculated potential harvests of bush-meat species in the markets of Bioko, Equatorial Guinea. Exploitation of some species could be much greater than is shown because not all animals killed are marketed. (Source: Fa *et al.* 1995.)

It is to be expected that markets will also influence hunter selectivity. A common observation in studies of subsistence hunters is that the more profitable animals are sold rather than being used for home consumption. It is possible, therefore, that hunting for cash, rather than food, will throw disproportionate pressure on to profitable species for as long as a particular selection strategy continues to pay off.

1.3   SUBSISTENCE USE AND HUMAN CARRYING CAPACITY

The concept of natural human carrying capacity, defined here as the population density of humans that can be supported by the natural productivity of a given area, can rarely be applied to existing communities. Very few people today try to survive without importing anything from outside their home range, and/or without increasing its human carrying capacity artificially – through agriculture. It is instructive, if only for comparative purposes, to take note of the sort of densities that might be sustainable when people are still living mainly by hunting and gathering for subsistence.

Clearly, tropical forest is far more supportive than the semi-deserts of Australia or the Kalahari, but at both extremes people evidently need to use the resources comprehensively by harvesting a wide range of plant and animal products, even though relatively few animals and plants may be in the category of first preference.

The manner is which human groups distribute themselves is also crucial to their survival. If they are nomadic, they can spread their harvesting effort continuously and avoid any local depletions. In 100 000 km^2 of sandhill desert country in central Australia, the population density of the Walbiri people was about 1 person to 90 km^2. Their neighbours, the Aranda, in a more productive territory of 65 000 km^2, numbered about 2000 – a density of around 1:30 km^2. In the more fertile Daly River region of Australia's Northern Territory, one small group lived at a density of 1 person to 13–18 km^2, while in north-east Arnhem Land the average density was somewhat less than 1 person to 20–23 km^2 (Kirk 1981). In Tasmania the pre-European population is estimated to have been about 3000–4000 people. That would work out at about 1 person to 10–15 km^2 because only two-thirds of Tasmania was inhabited (Jones 1974, cited in Kirk 1981). Estimates made by Silberbauer (1972) of G/Wi bushman bands in thornbush country of the Kalahari desert during the early 1960s averaged about 1 person to 11 km^2.

If long-term settlements are being established, the communities must be sufficiently far apart to allow adequate wildlife catchment areas in between. Thinking about natural carrying capacity in these terms is presumably second nature to hunter–gatherers who live without agriculture. Meggitt (1962) spent some time during the 1950s living among the Walbiri Aborigines of Australia's Northern Territory. He records that their daily food-gathering excursions extended within a radius of about 10–15 miles (18–27 km) of the permanent camps. Within that range there was little game to be seen. In the Kalahari desert of Africa, bands of G/Wi bushmen seldom spent more than three weeks in a camp. After that time food supplies would be exhaused within a 5-mile (8 km) radius and the band would move on (Silberbauer 1972).

A compromise between nomadism and a settled existence can work very well. Suriname still has a forested interior that is largely undisturbed. Rural communities practise a slash and burn form of shifting cultivation but are able to incorporate lengthy recovery (fallow) periods. Although serious local depletions of wildlife may occur as a result of hunting, the animal populations are given time to recover, along with the forest, after depleted areas are abandoned (Mittermeier 1991).

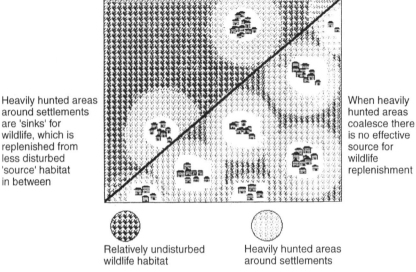

Heavily hunted areas around settlements are 'sinks' for wildlife, which is replenished from less disturbed 'source' habitat in between

When heavily hunted areas coalesce there is no effective source for wildlife replenishment

Relatively undisturbed wildlife habitat

Heavily hunted areas around settlements

Figure 1.3   The principle of sources and sinks in the subsistence use of wildlife.

1.3.1   A plan for people and wildlife

Ecological principles of spacing and density were applied in planning the Yuquí Ethnodevelopment Project in Bolivia (Stearman and Redford 1995). In 1992, by Presidential Decree, the Yuquí Indians (referred to in section 1.2.3) were given a homeland of 1150 km² called the Yuquí Indigenous Territory. Under the ethnodevelopment plan the population of about 150 Indians will be centred on 10 camps located strategically (to deter encroachment by settlers), in agreeable locations, and spaced so that each will have an adequate catchment for hunting. It is anticipated that around each camp an area of approximately 5–10 km radius, or the distance a forest hunter can exploit without staying out overnight, will be overhunted but it is expected that these zones will be replenished by immigration of wildlife from further away. Clearly this depends on having the camps so far apart that the land between can function as a source of wildlife because the overhunted zones will become sinks for dispersing animals (Figure 1.3).

The human population density in the Yuquí homeland (1 person to 6.6 km²) compares favourably with the estimated mean for lowland aboriginal people in forested Amazonia, which was about 1 person to 5 km² (Denevan 1976, cited in Vickers 1991). A density of 1:5 km² was also reported by Vickers (1991) in his study of the Siona-Secoya Indians in Equador. Vickers concluded that, at that density, the people were living 'well within the faunal resources offered by their environment' and that only one preferred species, the curassow (*Mitu salvini*), had been depleted through native hunting. However, within a zone of 590 km² around a village, there was also evidence of depletion of woolly monkeys and trumpeter birds (*Psophia crepitans*). Robinson and Redford (1991) suggested that the harvest rates for woolly monkeys, recorded by Vickers, were too high to be sustained even outside the 590-km² zone.

Evidently, it is not impossible for favoured species of animals to be overhunted, purely for subsistence purposes, even by people living in very productive environments, at low densities and in well spaced settlements.

1.3.2   Extinctions and subsistence

There is some evidence of local exterminations having been caused by preEuropean societies in the Pacific, either by hunting or forest clearance or by both (Pernetta and Hill 1986). There must have been easy hunting for people who colonized new areas and found animals which had never before encountered humans. Jared Diamond (1991) records being able to walk to within a few yards of the normally wary tree kangaroo (*Dendrolagus* sp.) in Papua New Guinea. This was only on Mount Foja, where there were no people. There is now a broad consensus that the giant moas of New Zealand were exterminated by a few thousand Maori colonists using only Stone Age weapons and fire (Jenkins 1992).

In general, however, it is evident that people *need* not destroy the natural environment by subsistence hunting and gathering provided they exist at low densities, use the resources comprehensively, and spread the harvest either by moving through the territory or by settling in small, well spaced communities. This combination of circumstances is now rare.

1.4   EFFECTS OF HUNTING TECHNOLOGY

In Papua New Guinea it has been remarked how readily hunters will chop down a forest tree in order to capture a single possum (Petauridae) or cuscus (*Phalanger* spp) (Pernetta and Hill 1986). Replacement of stone axes by steel ones has certainly reduced the work involved and presumably has affected judgements as to what size of tree might yield value for effort. Many groups have chain-saws which may similarly extend the sizes of clearings.

Most of the South American hunter–gatherers referred to in this chapter now use shotguns instead of traditional weapons. The same is true of the forest people in Papua New Guinea. Does this make much difference?

It is easier and quicker to kill forest animals with a blast of lead shot than with arrows, blowpipe darts or spears. Possibly, if a preferred species is so difficult to kill by traditional methods that demand cannot be met, then the introduction of shotguns may bring about heavier exploitation. This probably occurred with birds of paradise in Papua New Guinea. The decorative feathers are valued for ceremonial purposes but the birds live in forest canopy and are difficult targets. But it should not be assumed that forest people will always kill more animals if they have shotguns. Apparently, it depends on whether they feel it is worth killing more. It was thought likely that the Siwai people of Bougainville, Papua New Guinea, needed to spend less time hunting with shotguns than they had done with traditional weapons, but that they did not kill more game (Pernetta and Hill 1986).

In a South American study Alvard (1995) compared two 'hunter–horticulturalist' groups in Peru; one group using shotguns and the other group (which lived within the Manú National Park where shotguns were not permitted) using traditional weapons – primarily bows. Both groups harvested a comparable amount of meat per consumer per day but the group with shotguns did it in a quarter of the time. What they gained, therefore, was time rather than more meat. It was reasonable to conclude that meat consumption for both groups was felt to be adequate and there was no great incentive to kill more animals. But it was also concluded that more animals probably would be killed if more meat could be sold.

An important point in this context is the fact that the group with shotguns were actually hunting much more heavily (though not in per caput terms) as a result of their being a much bigger group; there were 250 of them compared with 100 using bows within the park. Moreover, because of access to basic western medicines, the larger group showed a much higher rate of population increase. At the reported rate of 4.7% p.a., they will double their numbers in 15 years. Local depletion of wildlife in the area has already been noted, especially for the large primates, specifically the black spider monkey (*Ateles paniscus*) and the red howling monkey (*Alouatta seniculus*).

1.5   IMPACT OF AGRICULTURE

The agricultural revolution actually proceeded very slowly, as evidenced by the thousands of years between sites of first appearance around the globe. For very long periods there were communities who practised shifting cultivation or very limited agriculture around small and scattered settlements. They used fire to maintain grasslands and, no doubt, they caused local extinctions as do people in such circumstances today. Some of this impact altered the landscape and faunal composition such that present-day landscapes may still reflect Neolithic human activity. This had led some writers to conclude that there is no such thing as virgin forest. However, at the Neolithic scale of operations it is quite possible that the human impact made a positive contribution to biological diversity. A shifting mosaic of clearings, regrowth and grassland–forest edge within a forest matrix is a natural feature of forest dynamics.

This last point is sometimes made, in the context of present-day conservation, to justify human activities within protected areas. But scale is all important: not just the extent of proposed activities within an area, but the spatial and temporal scale of the mosaic on a regional basis (Harris and Silva-Lopez 1992; Noss 1992). Over a large proportion of today's world there are no longer patches of agriculture in a matrix of forest; there are forest fragments in a matrix of agriculture. The distribution of faunal sinks and sources, referred to above, has been entirely reversed so that the basic principle, that use will be followed by replenishment from the wild, is no longer applicable. Present circumstances, for utilization and conservation, are in stark contrast with those of ancient times.

1.6 TECHNOLOGY, URBANIZATION AND CULTURAL CHANGE

There has been nothing slow about the spread of technology. It took millennia to get from the wheel to the microchip but only a few years for the microchip to cross the same cultural distance. People whose parents had only Stone Age technology are now using computers and satellite dishes. In South American forests, monkeys are simultaneously being collected for their meat and for the latest in biomedical research. Between these extremes people everywhere are in transition.

Technological change has both direct and indirect impacts on wildlife and conservation strategy. Motor vehicles, bulldozers and the spread of concrete need no further comment but the indirect effects – operating through culture and traditions – are less obvious.

It is sometimes assumed that people who are still living as hunter–gatherers, or who have recent connections with that way of life, will have the least environmental impact under modern circumstances. There is no hard evidence to support this belief and it tends to be used in political rather than scientific contexts (Redford 1990; Redford and Stearman 1993). Aboriginal peoples, who have survived in the wild, may have a profound knowledge of local plants and animals, but this does not necessarily make them more successful managers of social and ecological change.

And change is inevitable when traditional ways no longer meet growing needs and aspirations. In a current land dispute between Aboriginal groups in Australia, the Ngurrumunga clan is claiming rights, under the Native Titles Act, to land which is occupied by part of a silica mine. On national television, a spokesman for the clan said he felt 'great' at the prospect of having rights to mining royalties, and he expressed sympathy with those who only had national parks on their land (ABC 1995).

It may be saddening to see the spread of consumerism and the loss of cultural diversity but 'romancing the Stone Age' is not helpful to present-day conservation. As Redford and Stearman (1993) have put it: 'To expect indigenous people to retain traditional, low-impact patterns of resource use is to deny them the right to grow and change in ways compatible with the rest of humanity.' Yet the myth of the 'ecologically noble savage' (Redford 1990) does seem to influence some decision-makers. Throughout most of Canada and in Alaska, claims for subsistence hunting rights are complicating effective wildlife management because indigenous people are allowed to hunt without restriction under the Natural Resources Transfer Agreement of 1930 (Regelin 1991).

The point to be taken is that the impact of human beings on the environment depends not on their ethnic identity, but on what they do, and in what numbers. Indigenous peoples and conservation biologists can both contribute to the effort of trying to maintain biological diversity, but there is no reason to assume that the ownership and occupation of an area by indigenous people will be enough to guarantee the security of its ecological integrity. If only conservation problems were so easily solved!

It is probably more realistic to accept that the majority of people everywhere have come to regard native fauna, not as part of the local eco-complex that was once the human environment, but as a collection of remaining wild things that are to be selectively treasured, tolerated, harvested, disregarded, or exterminated according to received wisdom and prevailing attitudes. It is mostly in this context that we shall consider the conservation and utilization of wildlife in the modern world.

REFERENCES

ABC (1995) *Four Corners: Title Fight*, TV Documentary broadcast on 9 October 1995. Australian Broadcasting Corporation, Sydney.

Alvard, M. (1993) Testing the 'Ecologically Noble Savage' hypothesis: interspecific prey choice by Piro hunters of Amazonian Peru. *Human Ecology*, 21 (4), 355–87.

Alvard, M. (1995) Shotguns and sustainable hunting in the Neotropics. *Oryx*, 29 (1), 58–66.

Asibey, E.O.A. and Child, G.S. (1990) Wildlife management for rural development in sub-Saharan Africa. *Unasylva*, 161 (41), 3–10. FAO, Rome.

Borshay, R. (1972) The !Kung Bushmen of Botswana, in *Hunters and Gatherers Today* (ed. M.G. Bicchieri). Holt, Rinehart and Winston, New York, pp. 327–68.

Denevan, W.M. (ed.) (1976) The aboriginal population of Amazonia, in *The Native Population of the Americas in 1492*. Wisconsin Press, Madison, Wisconsin, pp. 205–34.

Diamond, J. (1991) *The Rise and Fall of the Third Chimpanzee*. Radius, London.

Fa, J.E., Juste, J., Del Val, J.P. and Castroviejo, J. (1995) Impact of market hunting on mammal species in Equatorial Guinea. *Conservation Biology*, 9 (5), 1107–15.

Fitzgibbon, C.D., Mogaka, H. and Fanshawe, J.H. (1995) Subsistence hunting in Arabuko-Sokoke Forest, Kenya, and its effects on mammal populations. *Conservation Biology*, 9 (5), 1116–26.

Hames, R. (1991) Wildlife conservation in tribal societies, in *Biodiversity: Culture, Conservation, and Ecodevelopment*. Westview Press, Boulder, Colorado, pp. 172–99.

Harcourt, A.H. and Stewart, K.J. (1980) Gorilla-eaters of Gabon. *Oryx*, **xv** (3), 248–51.

Harris, L.D. and Silva-Lopez, G.S. (1992) Forest fragmentation and the conservation of biological diversity, in *Conservation Biology: The Theory and Practice of Nature Conservation Preservation and Management* (eds P.L. Fiedler and S.K. Jain). Chapman & Hall, London, pp. 197–237.

Jenkins, M. (1992) Species extinction, in *Global Biodiversity: Status of the Earth's Living Resources* (compiled by Wildlife Conservation Monitoring Centre). Chapman & Hall, London, pp. 192–205.

Jones, R. (1974) Tasmanian tribes, in *Aboriginal Tribes of Australia* (ed. N.B. Tindale). Australian National University Press, Canberra.

Keyfitz, N. (1989) The growing human population. *Scientific American*, **261** (3), 70–7.

King, S. (1994) Utilisation of wildlife in Bakossiland, West Cameroon, with particular reference to primates. *Traffic Bulletin*, **14** (2), 63–73.

Kirk, R.L. (1981) *Aboriginal Man Adapting: The Human Biology of Australian Aborigines*. Clarendon Press, Oxford.

Meehan, B. (1977) Hunters by the seashore. *Journal of Human Evolution*, **6**, 363–70.

Meggitt, M.J. (1962) *Desert People: A Study of the Walbiri Aborigines of Central Australia*. Angus and Robertson, Sydney.

Mittermeier, R.A. (1991) Hunting and its effect on wild primate populations in Suriname, in *Neotropical Wildlife Use and Conservation* (eds J.G. Robinson and K.H. Redford). University of Chicago Press, Chicago, pp. 93–107.

Noss, R.F. (1992) Issues of scale in conservation biology, in *Conservation Biology: The Theory and Practice of Nature Conservation, Preservation and Management*. Chapman & Hall, London, pp. 239–50.

Peres, C.A. (1991) Humboldt's woolly monkeys decimated by hunting in Amazonia. *Oryx*, **25** (2), 89–95.

Pernetta, J.C. and Hill, L. (1986) The impact of traditional harvesting on endangered species: the Papua New Guinea experience, in *Endangered Species: Social, Scientific, Economic and Legal Aspects in Australia and the South Pacific*. Total Environment Centre, Sydney, pp. 98–130.

Redford, K.H. (1990) The ecologically noble savage. *Orian*, **9**, 24–9.

Redford, K.H. and Robinson, J.G. (1991) Subsistence and commercial uses of wildlife in Latin America, in *Neotropical Wildlife Use and Conservation* (eds J.G. Robinson and K.H. Redford). University of Chicago Press, Chicago, pp. 6–23.

Redford, K.H. and Stearman, A.M. (1993) Forest-dwelling native Amazonians and the conservation of biodiversity: interests in common or in collision? *Conservation Biology*, **7** (2), 248–55.

Regelin, W.L. (1991) Wildlife management in Canada and the United States, in *Global Trends in Wildlife Management* (eds B. Bobek, K. Perzanowski and W. Regelin). Trans. 18th IUGB Congress, Krakow, 1987. Swiat Press, Krakow–Warszawa.

Robinson, J.G. and Redford, K.H. (1991) Sustainable harvest of Neotropical forest animals, in *Neotropical Wildlife Use and Conservation*. (eds J.G. Robinson and K.H. Redford) University of Chicago Press, Chicago, pp. 415–29.

Silberbauer, G.B. (1972) The G/Wi Bushmen, in *Hunters and Gatherers Today* (ed. M.G. Bicchieri). Holt, Rinehart and Winston, New York, pp. 271–326.

Stearman, A.M. and Redford, K.H. (1995) Game management and cultural survival: the Yuquí Ethnodevelopment Project in lowland Bolivia. *Oryx*, **29** (1), 29–34.

Vickers, W.T. (1991) Hunting yields and game composition over ten years in an Amazon Indian territory, in *Neotropical Wildlife Use and Conservation* (eds J.G. Robinson and K.H. Redford). University of Chicago Press, Chicago, pp. 53–81.

2

Conservation and use of wildlife in the modern world

2.1 THE IMPACT OF HUMAN NUMBERS

Hundreds of millions of people are still heavily dependent on 'bush foods' for subsistence but are no longer using them sustainably. In the Indian sub-continent people now live at an average density of 236 people per km^2, more than a thousand times the density of those in rural Amazonia. The remaining forest habitats of India are constantly subjected to a virtual scouring by livestock and people in search of 'minor forest produce' (Dang 1991; Panwar 1992). The situation is similar in some other parts of Asia and in parts of Africa.

Accepting that at the beginning of the Neolithic there were between 5 and 10 million people in the world – call it 7.5 million – then, during 1995, the equivalent of the entire Neolithic population was added to the world's total every single *month* – an increase of 90 million people for the year. Rapid population growth is a recent phenomenon. There were fewer than 0.5 billion people on Earth during the seventeenth century and not more than about 1.5 billion at the beginning of the present century. The steepest part of the growth curve dates from about 1920, when mortality rates around the world began to yield to modern medicine. The rate of increase has fallen slightly in recent years, an effect of the demographic transition resulting from improved living standards, but because the rate of increase applies to an ever-larger base, the growth curve continues steeply upwards (Figure 2.1). In the most likely scenario the world population will stabilize shortly after the year 2200 at about 11.6 billion, most of the increase occurring in the developing world (UN 1992).

Human numbers are at the root of most global conservation problems. The inequitable use of resources among nations does not counter that conclusion. Some reduction in environmental impact can be expected as and

Conservation and the Use of Wildlife Resources. Edited by M. Bolton.
Published in 1997 by Chapman & Hall. ISBN 0 412 71350 0.

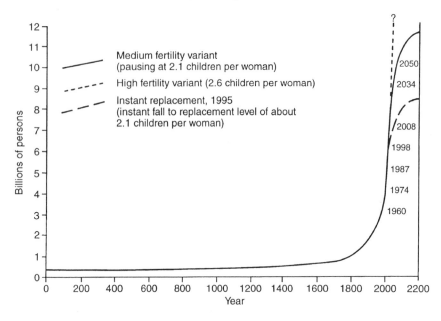

Figure 2.1 World population from year 0 to stabilization. The medium fertility variant is considered the most likely scenario and will result in a stable population of 11.6 billion shortly after the year 2200. The high fertility variant would produce a population (still growing if that were possible) of 28 billion by the year 2150. By that year, if fertility had continued as at present and mortality had not increased (an impossibility), the population would be over 600 billion. Note that even if fertility had dropped instantly to replacement level in 1995, the population, because of its structure, would continue to grow until the year 2150, when it would stabilize at 8.4 billion. (Source: UN Population Division.)

when wealthier nations adopt a leaner lifestyle and cleaner technology but any realistic reductions will be heavily outweighed by modest and essential improvements in the living standards of those in developing countries. Consider the two passages below :

> Those who are poor and hungry will often destroy their immediate environment in order to survive. They will cut down forests; their livestock will overgraze grasslands; they will overuse marginal land; and in growing numbers they will crowd into congested cities. The cumulative effect of these changes is so far-reaching as to make poverty itself a major global scourge.
>
> (World Commission on Environment and Development 1987)

Current projection sees a doubling of human population in the next 50 years. To bring people in the developing world to the same standard of living as people in the developed world would require a 500% increase in world GDP. Doubling the population as well would require a ten-fold increase. To do this without increasing pollution (and the assimilative capacity of our ecosystems are already stretched) would require a 90% reduction in pollution per head of population.

(Passage produced by the Johnstone Centre 1995, from Young (UNESCO) 1992)

The crux of the problem, in environmental terms, is that both passages must be taken seriously.

Pollution is, of course, only a by-product, an indirect measure, of land and resource use. The population problem is emphasized here because of the basic needs of so many people; the rich, the poor and those in every stage between, translate into direct demands on natural habitats and the living space of other species. Agriculture and urbanization would remain major threats to wildlife even with no pollution. In a number of countries conservation plans must allow for another doubling of population within the next 20–30 years.

Reflecting on the fact that, with double the present numbers, human beings may be appropriating more than 80% of the Earth's produce, Swanson (1992) has flatly stated that

The prospects for sufficient resources being made available for all of these human needs and also for those of other species is zero . . . Difficult choices will have to be made, and there will inevitably be a much lower number of species; the question is not whether such a reduction will occur, but whether it will be conducted out of complete ignorance and neglect, or not.

United Nations population projections indicate that the amount of arable land per person will fall to an average of only 0.16 ha within the next 30 years (Engelman 1995). The pressure to cultivate more land – including the existing protected areas – is expected to intensify. Global trade distributes the pressures; Australia's less populous but fertile zones, for example, feed some 60 million people in other countries. Certainly there can be no backing away from intensive agriculture without bringing more land into production.

In a few western countries populations have stabilized or are even showing signs of decline. New economies based on workers and consumers with a higher average age will have to be developed for the future but there is little sign of this happening. Indeed, it is revealing that the word 'stagnation', was used by *The Economist* (12 March 1994) in forecasting population stability for Russia, the Czech Republic and Portugal.

2.2   THREATS AND RESPONSES

2.2.1   Swift paths to extinction

The extinction of a species represents a fundamental and irreversible loss to biological diversity – precluding any further evolutionary change in that species' lineage.

It is quite possible for deliberate persecution or exploitation to be the main threat to a species, and a number of large mammals and birds have been exterminated or are endangered in this way. The plight of the world's rhinos is a familiar example, as is the history of whaling. But by far the most serious threat, globally, is the destruction and modification of habitats, mainly by agriculture and settlement. The destruction of tropical forests has been extensively documented and publicized but other vegetation types are being destroyed at an even faster rate. Regelin (1991) draws attention to the particularly rapid destruction of wetlands in North America. California has already lost 90% of its wetlands and the losses in several southeastern states have approached 80% in recent years. Drainage for agriculture has been responsible for 87% of the loss while 8% has been the direct result of urbanization.

Among currently threatened species of mammals and birds, habitat loss affects 76% and 60% respectively. Direct exploitation (mainly for meat) affects half the threatened species of mammals and 29% of birds. Introduced predators and competitors present another major problem (Jenkins 1992).

More often than not, different types of threat operate simultaneously and the effects may compound each other. Clearance of cover, for example, may leave animals or their nests more vulnerable to predation. Habitat fragmentation which separates animal populations into smaller sub-populations can start off the demographic and genetic downward spirals which have come to be known as extinction vortices (Gilpin and Soulé 1986; Wilcove *et al.* 1986; Soulé 1987) Certain life-history characteristics, such as ability to disperse, degree of specialization, reproductive potential, and longevity, are likely to determine an animal's sensitivity to habitat fragmentation (Karr 1991; Lawrence 1991; Jenkins 1992; Pearl 1992; see also Chapter 3, section 3.1.3).

2.2.2   The legislation response: species and habitats

Historically, national governments have most commonly sought to protect wild animals through legislation which sets different categories of protection and then lists the species in each category. In the United Kingdom, for example, the Protection of Birds Act of 1954 listed gamebirds separately and assigned other species to Schedules according to whether they were fully protected; could be 'taken' outside a closed season or could be taken at any time (e.g. wood pigeon). This type of legislation is used for mammals and birds in most countries. Where it is adequately enforced it can be useful to protect those species which are primarily threatened by direct exploitation, but it fails totally to protect the habitat of even the rarest or most endangered species.

The US Endangered Species Act of 1973 represents an extension to the 'listed species' type of legislation. Under this Act there is provision for protecting 'critical habitats' of endangered species where such areas are defined as being essential to the species' conservation. The areas must be designated and described in the *Federal Register*. The US Endangered Species Act was unusual in that it linked species protection with habitat even in 'inconvenient' locations and it stipulated that no activity by a government agency should lead to the extinction of a listed species. Even so, it has been pointed out that habit protection under the Act is still incomplete for species listed as endangered or threatened (Sidle and Bowman 1988). The much-publicized case of the northern spotted owl (*Strix occidentalis caurina*) demonstrates the increasing difficulty of protecting wildlife habitat for particular species. The owl survives only in mature and old-growth conifer forest of the Pacific Northwest and the needs of the owl are now in conflict with the demands of a logging industry which is the economic mainstay of the region (Harrison *et al*. 1993).

2.2.3   The area protection response

During the latter half of this century governments have increasingly adopted a habitat approach to conservation. The common response has been to select areas for protection and administration as national parks, nature reserves, wildlife sanctuaries or some other legally defined category of protected area. Over 80% of the world's protected areas have been established since 1962 (WCMC 1992). The World Conservation Union (IUCN) developed a widely adopted classification system for protected areas based on management objectives (Box 2.1). However, terminology, legal mechanisms, levels of funding and standards of enforcement can vary significantly between nations so that international comparisons can be misleading.

The World Conservation Monitoring Centre (WCMC 1992), following the criteria used in compiling the United Nations List of National Parks and Protected Areas (IUCN 1990), has calculated that only about 3% of the land's surface can be considered to be totally protected, and that includes the enormous contribution made by the Greenland National Park. If partially protected areas are included (all areas of 1000 ha or more which conform to Categories 1–V of the UN list), then protected land amounts to slightly over 5% (Table 2.1). The World Commission on Environment and Development (WCED 1987) has stated that 'the total expanse of protected

---

Box 2.1 IUCN categories and management objectives of protected areas

I *Scientific Reserve/Strict Nature Reserve*: to protect nature and maintain natural processes in an undisturbed state in order to have ecologically representative examples of the natural environment available for scientific study, environmental monitoring, education, and for the maintenance of genetic resources in a dynamic and evolutionary state

II *National Park*: to protect natural and scenic areas of national or international significance for scientific, educational and recreational use

III *Natural Monument/Natural Landmark*: to protect and preserve nationally significant natural features because of their special interest or unique characteristics

IV *Managed Nature Reserve/Wildlife Sanctuary*: to assure the natural conditions necessary to protect nationally significant species, groups of species, biotic communities, or physical features of the environment where these require specific human manipulation for their perpetuation

V *Protected Landscape or Seascape*: to maintain nationally significant natural landscapes which are characteristic of the harmonious interaction of man and land while providing opportunities for public enjoyment through recreation and tourism within the normal lifestyle and economic activity of these areas

Other categories defined by IUCN are: VI (*Resource Reserve*), VII (*Natural Biotic Area/Anthropological Reserve*) and VIII (*Multiple Use Area/Managed Resource Area*)

Source: This abridged list is from WCMC (1992).

---

areas needs to be at least tripled if it is to constitute a representative sample of the Earth's ecosystems'.

Assuming that protected areas are secure and effectively managed, they will still only conserve a nation's biodiversity to the extent that the protected areas are all in the right places. And in order to ensure that viable examples of all ecosystems are included in a national network, an ecologically diverse country will need to have more separate areas than an ecologically homogeneous nation. In practice, protected areas are not often distributed in a scientifically planned network at all. A great many factors can influence the selection of sites and sometimes land is declared a 'nature reserve' simply because it is not wanted for other purposes.

Table 2.1 Distribution and coverage of protected areas by biome (from Johnston 1992. Biome definitions after Udvardy 1975.)

Biome type	Protected areas		Biome area	
	Number	Area (km²)	(km²)	% Total area
Subtropical/temperate rainforests/woodlands	935	366 100	3 928 000	9.32
Mixed mountain systems	1 265	819 600	10 633 000	7.71
Mixed island systems	501	246 300	3 244 000	7.59
Tundra communities	81	1 643 400	22 017 000	7.46
Tropical humid forests	501	522 000	10 513 000	4.96
Tropical dry forests/woodlands	807	818 300	17 313 000	4.73
Evergreen sclerophyllous forests	786	177 400	3 757 000	4.72
Tropical grasslands/savannas	56	198 200	4 265 000	4.65
Warm deserts/semi-deserts	296	957 700	24 280 000	3.94
Cold-winter deserts	139	364 700	9 250 000	3.94
Temperate broadleaf forests	1 509	357 000	11 249 000	3.17
Temperate needle-leaf forests/woodlands	440	487 000	17 026 000	2.86
Lake systems	18	6 600	518 000	1.28
Temperate grasslands	196	70 000	8 977 000	0.78
Classification unknown	961	700 800	–	NA
Totals	8 491	7 735 100	146 970 000	5.26

Many valuable areas fail to qualify for the UN list. These include sites which are too small and/or not managed by the 'highest competent authority'. They may be managed by local councils, hunters or the military, or they may be protected by virtue of superstition, religion or just isolation, but they conserve significant amounts of biodiversity (Johnston 1992).

Protected areas, then, despite the inadequacies of global coverage, are the mainstay of habitat conservation.

2.2.4 International responses

Numerous multilateral treaties with global or regional scope have some relevance to biodiversity conservation. A few of them are widely believed to exert a strong influence on conservation policy in member states; these include the Wetlands Convention (Ramsar, 1971); the World Heritage

Convention (Paris, 1972), and the Convention on International Trade in Endangered Species, known by its acronym, CITES (Washington, 1973).

There are high hopes for the Convention on Biodiversity (Rio, 1992) and the Law of the Sea (Montego Bay, 1982) but these only came into force in 1993 and 1994 respectively.

2.2.5   CITES

Although there were only 21 initial signatory nations to CITES (which entered into force in July 1975), the Convention quickly became the largest international wildlife conservation treaty and there are currently 130 nations pledged to implement its provisions. CITES is exclusively concerned with international trade, not domestic use of wildlife, and it shows both the strengths and weaknesses of international conservation efforts.

(a) How CITES works

CITES came into force in July 1975. As a core function the Parties to the Convention are obliged to monitor the international trade in wild fauna and flora and take action to safeguard any species whose survival is threatened by trade. Taxa which are selected for CITES protection are listed in one of three Appendices.

Appendix I
Species threatened with extinction that are, or may be, affected by trade. With narrow exceptions, mainly for zoological and scientific purposes, all international trade in Appendix I species is prohibited. There are currently about 675 listed species, including all rhinos, great apes, sea turtles and most big cats.

Appendix II
Species not yet threatened by extinction but strict control of international trade is considered essential to prevent a decline to that status. More than 25 000 species have been listed and over 21 000 of them are plants. The Convention allows commercial trade in these species under a permit system.

Appendix III
A special category in which any Party may list species of particular national concern. Canada, for example, has listed the walrus (*Odobenus rosmarus*).

CITES is financed primarily from member contributions which are based on the UN contribution scale. This ranges from a few hundred dollars for the poorest countries to over a million dollars for the richest. The USA provides about 25% of the CITES budget. CITES has a permanent secretariat with full-time staff in Geneva, but amendments to the appendices and policy issues are decided by the CITES party nations. These meet every second year in a Conference of the Parties (COP), each of which is hosted by a different party nation.

Non-government organizations (NGOs) representing conservationists, scientists and industry interests around the world have been extremely active as non-voting participants in support of CITES. The treaty specifically allows for their involvement.

(b) Problems with CITES
Although CITES has had outstanding success in securing international cooperation, much still depends on the political will of the party nations because the CITES Secretariat has very limited authority. On becoming a party to CITES, a nation agrees to set up management and scientific authorities through which to implement and enforce the treaty's policies and decisions. Implementation and enforcement, therefore, ultimately depend on the capacity and commitment of member nations to comply with the treaty's requirements. There is much variation in this regard; some parties of long standing have not even enacted legislation through which the CITES mandates could be enforced (Hemley 1994).

Finance has been another persistent problem and in recent years as many as one-third of the parties have failed to make their agreed contributions (Hemley 1994).

An even more fundamental problem with the treaty derives from the fact that it was originally intended to protect those species, such as crocodiles and the spotted cats, which were conspicuously threatened by trade. But for most species trade is not the major threat and it has frequently been argued, especially by, or on behalf of, developing countries, that a total prohibition of trade could work against the conservation of some species by precluding trade which might earn revenue for conservation purposes (e.g. Swanson 1992). This begs the question, of course, that revenue from wildlife trade *would be used* for conservation.

The criteria by which species are selected for listing in the appendices have also been heavily criticized. Critics say the criteria, often referred to as the 'Berne criteria' because they were adopted at the first COP in Berne (1976), are not sufficiently objective.

At the ninth COP, in Florida (1994), a resolution was adopted, repealing the Berne criteria and setting out new ones for the amendment of Appendices I and II. The problem of obtaining accurate and up-to-date information for a great many species, however, will not be easily solved.

(c) Ranching and the quota system
In response to the argument from developing countries that possible economic benefits from trade should be realized, CITES moved towards greater flexibility. At the third COP (New Delhi, 1981) a resolution (Conf. 3.15) was adopted to provide for populations of Appendix I species to be

downlisted to Appendix II for the purpose of ranching, defined as the rearing in a controlled environment of specimens taken from the wild. For CITES approval, a ranching scheme should be expected to benefit the local (ranched) population and products of the operation must be adequately identified and distinguishable from products of Appendix I populations.

The first ranching proposal to be accepted resulted in Zimbabwe's Nile crocodile (*Crocodylus niloticus*) population being downlisted from Appendix I to Appendix II (Chapter 7). New and separate guidelines have recently been adopted for evaluating marine turtle ranching proposals (Florida, 1994, Conf. 9.20).

Further reform in the direction of trade regulation, rather than prohibition, was adopted at the fourth COP (Gaberone, 1983) when it was accepted that a number of leopard skins could be traded to 'enhance the survival of the species' (Conf. 4.13). The quota was set at 450 skins, shared between states within the leopard's range. The quota was successively increased and was set at 2055 in 1992 (Conf. 8.10). Quotas have since been used in combination with ranching proposals.

The task of CITES has become far more complicated than was originally foreseen. It is extremely difficult to control international trade in wildlife products – especially when some of the products can scarcely be identified. It would be much easier to enforce a total ban. But because trade is not the major threat to most species, a total ban would not necessarily help. For some species, in some countries, it could make matters worse. For this reason the outlook for CITES is one of increasing complexity. The full text of the Convention can be found in Hemley (1994).

2.3   CONSERVATION THROUGH UTILIZATION

2.3.1   Purposes of use

The many material purposes for which wild animals are valued can be considered in five categories:

- food and medicinal products
- non-edible products (skins, fur, feathers, horns, bones, teeth, shells, bait for other species, etc.)
- live animals (pet trade, zoos, biomedical research, working animals, domestication)
- sport (hunting, fishing, racing)
- tourist attractions (non-consumptive use)

The only entirely new category since prehistoric times is that of tourism, though the range of live uses and the use of animals for sport have certainly burgeoned from very small beginnings. Many species have multiple uses or provide a variety of useful products in one or both produce categories.

Deer, for example, can have value in all five categories with a range of products including meat, skins, antlers and velvet. At the other extreme, some species are valued only for a single body part or substance.

The values are not constant; everywhere from caves to fashion houses they are subject to the laws of supply and demand and will be influenced by availability of alternatives and changing traditions.

In this book there are examples of utilization strategies for all categories of purpose, involving multiple uses and a wide range of taxa.

2.3.2 Levels of use

Three broad levels can be distinguished:

- consumer level use only
- exploitation for local markets
- exploitation for wider markets and/or export

The three levels are obviously not mutually exclusive; it is common for subsistence hunters to sell some meat at the local market, though it is less common for the same hunters to be involved in wider commerce. The importance of the commercial factor cannot be overemphasized. To the extent that utilization is commercialized in conservation schemes, the schemes may be subject to all the economic complexity and uncertainty of modern markets.

2.3.3 Forms of use and the problems of specialization

We have seen that the subsistence hunter–gatherer harvests a broad spectrum of wildlife, selecting from whatever nature has to offer. In contrast, the operator who supplies markets is selecting for profitability and trying to meet off-site demands which frequently cannot be met from the natural productivity of an area. Selectivity becomes specialization and the focus then becomes centered on production of the selected species. Most terrestrial utilization schemes are necessarily based on single species or a very few closely related taxa.

Management interventions to increase productivity of particular species often also affect the ecology of animals and plants which are not being favoured. The impact may be profound if habit change is involved because populations are more vulnerable to a manipulation of their habitat than they are to a direct manipulation of their numbers (Caughley 1977) The end result, intended or otherwise, may be a degraded ecosystem and lower biological diversity. Domestic animals and plants are the extreme example: only a few species are involved yet vast tracts of the Earth have been converted and are maintained for their production. In the language of economics, the resource base has been homogenized and segregated in accordance with the law of specialization (Luxmore and Swanson 1992).

Apparently, there is a dilemma: if selective productivity is destructive of natural systems, how can wildlife utilization be turned to conservation advantage?

2.3.4   Damage limitations and conservation gains

Possible interventions for increasing productivity of a species include:

- increasing the supply of a limiting factor at critical times (water holes, winter feed, nesting sites)
- modification of general habitat to favour the target species (planting, clearing, burning, diversification, leaving uncultivated patches)
- destruction/removal of competing species, including predators
- protection of target species at critical times (turtle nests, colony-nesting birds)
- captive breeding or rearing and release

This is quite a broad range of management possibilities and the degree of environmental impact in a utilization scheme will very much depend on which intervention, or combination of them, is required. For example, there need be very little ecological disturbance in protecting a turtle-nesting beach so that a proportion of the eggs can be collected. It depends what the beach is being protected from.

There is a second, related, reason why the productivity–specialization dilemma does not always preclude conservation advantage in utilization schemes: in practice, conservation value can only be meaningfully assessed if the alternatives to utilization are also considered. A planned utilization scheme might be a lesser evil, in environmental terms, than the status quo, or any number of other probabilities. Those who are in principle opposed to using wildlife commonly argue as though the alternative to utilization is to allow wildlife to remain undisturbed. They have a powerful argument when that is the case. Unfortunately, because of the pervasiveness of the threats, it is usually not the case and decision-makers should take into account the predicted consequences of not using the resource, as well as the consequences of using it in the way proposed. The concept is illustrated with a simple Cartesian plane in Figure 2.2.

Finally, there is a third factor to be considered: involvement with wildlife that is constructive (again, relative to real alternatives and not imagined ideals) may be indirectly beneficial if it provides money, goods or services (e.g. research opportunities) which can serve conservation objectives in the longer term.

There can be no doubt that management for selective production is often incompatible with the conservation of biodiversity. But this is a constraint, not a total block on the inclusion of wildlife utilization in the conservation armoury. Nor is it always the major constraint; other limitations may

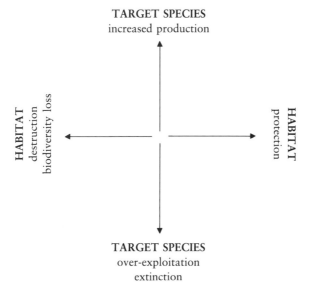

Figure 2.2 The possible relationships between wildlife utilization and conservation can be visualized at an infinite number of points on a two-dimensional plane.

operate when the ecological requirements are well understood and adequately met − as exemplified by the use of shearwaters in Tasmania.

2.4 HARVESTING MUTTONBIRDS

2.4.1 Biology

The short-tailed shearwater (*Puffinus tenuirostris*), or Tasmanian muttonbird, is a migratory seabird which nests on islands and shores of southeastern Australia. The biggest colonies are around Tasmania, where over 9 million pairs nest in at least 167 colonies. Major concentrations of these occur on islands in Bass Strait. The biology of muttonbirds has been studied since 1947 by the Commonwealth Scientific and Industrial Research Organization (CSIRO) and Tasmanian Government scientists. The details that follow are from Skira (1992) and Ramsay (1994).

The birds lay a single egg in a burrow and are capable of breeding from the age of 4 until 15 years. Mean age at first breeding is around 7 years and mean survival time is 9.3 years after first breeding, but not all birds breed. From a sample of 418 birds, 27% produced no young during their lives and 19% had only one laying. Colonies are largely composed of the offspring of a relatively small proportion of reproductively successful birds.

2.4.2   Exploitation

Adult muttonbirds are fully protected, but under Tasmanian state law, chicks may be commercially harvested between 27 March and 30 April each year. Maximum sustainable yield is estimated at 37% which amounts to about 1.63 million chicks. Non-commercial catchers are separately licensed to take birds for personal consumption but not for sale.

Traditional exploitation practices have not changed in more than 150 years. Tasmanian Aborigines have a particular association with the industry and very few non-Aborigines are now commercially involved. During the short open season, chicks, aged about 10 weeks, are dragged from their burrows by licensed catchers, their necks are broken and they are hung in rows on a spit threaded through their lower mandibles. In this way dozens of birds can be carried by a catcher to the processing shed, operated by licensed processors. In the sheds the birds are plucked and eviscerated and oil from the proventriculus is drained into a drum. The carcases are scalded and allowed to cool on racks. The processing sheds are subject to inspection and approval by health authorities.

Most meat is sold through retail outlets in Tasmania, though a substantial proportion is exported to New Zealand. In 1990, a total of 68 284 birds, representing 33.7% of the commercial harvest, was exported. The meat is oily and has a distinctive flavour. It is not considered to be a gourmet product and the end-users are mostly older people, especially Aborigines.

Feathers, though of low value, are sold within Australia for bedding. The proventricular oil is sold mostly to the local racehorse industry as a food additive to improve the coat of horses. It is also used as a leather dressing. Gross income from meat, feathers and oil in 1990 amounted to A$260 000 – an average return of only A$1.28 (£0.60) per bird.

2.4.3   Characteristics of the muttonbird industry

These can be summarized as follows:

- The resource is large and remarkably dependable.
- No active management of the resource is required.
- The bird colonies, identified as the resource, are clearly defined, scientifically quantified and monitored.
- Exploitation procedures are officially and responsibly regulated.
- Offtake is well within the sustainable yield of the resource so the harvest poses no threat to the wild population.
- The island nesting-grounds are not disturbed by livestock (which trample the burrows and compact the ground), partly because the value of the industry has been a major influence on landowners. The industry is

therefore considered to have been beneficial in conservation terms (Serventy 1966).

- Although the industry is ecologically sound, it is nevertheless in decline. In recent years the harvest has been less than a third of the permitted quota. The reasons for this are commercial and socio-cultural.
- As a tradition, the muttonbird harvest was an important social event for Aborigines. It is now much less important, especially among younger people.
- Commercial problems include increased marketing costs and lower demand for the product. Rejuvenation of the industry depends on broadening the markets and perhaps developing new niche markets in Asia and the Pacific. However, the industry does not have the infrastructure or skills base for these tasks.

The muttonbird industry, then, is an example of a single-species, simply structured industry, operating at all three levels (consumer, local market and export), with conservation benefit gained through protection of nesting habitat. It is ecologically and administratively secure but is in decline for socio-economic and socio-cultural reasons. Other wildlife utilization projects will be seen to differ fundamentally in similar simple analyses.

ACKNOWLEDGEMENTS

I am grateful to Nada Chaya of Population Action International for information on human population growth. I am also pleased to thank Judy Iland, Brenda Miller and Leonie Wyld for their fact-finding on my behalf.

REFERENCES

Caughley, G. (1977) *Analysis of Vertebrate Populations*. John Wiley & Sons, Chichester.

Dang, H. (1991) *Human Conflict in Conservation: Protected Areas: the Indian Experience*. Har-Anand Publications, New Delhi.

Engelman, R. (1995) Feeding tomorrow's people from today's land. *Environmental Conservation*, 2 (22), 97–8.

Gilpin, M.E. and Soulé, M.E (1986) Minimum viable populations: processes of species extinction, in *Conservation Biology, the Science of Scarcity and Diversity* (ed. M.E. Soulé). Sinauer, Sunderland, Mass., pp. 19–34.

Harrison, S., Stahl, A. and Doak, D. (1993) Spatial models and spotted owls: exploring some biological issues behind recent events. *Conservation Biology*, 7 (4), 950–3.

Hemley, G. (ed.) (1994) *International Wildlife Trade: A CITES Sourcebook*. WWF and Island Press, Washington, DC.

IUCN (1990) *1990 United Nations List of National Parks and Protected Areas*. IUCN, Gland.

Jenkins, M. (1992) Species extinction, in *Global Biodiversity: Status of the Earth's*

Living Resources (compiled by Wildlife Conservation Monitoring Centre). Chapman & Hall, London, pp. 192–205.

Johnston, S. (1992) Protected areas, in *Global Biodiversity: Status of the Earth's Living Resources* (compiled by Wildlife Conservation Monitoring Centre). Chapman & Hall, London, pp. 447–78.

Johnstone Centre (1995) Johnstone Centre of Parks, Recreation and Heritage: An Interdisciplinary Centre of Research, Consultancy and Teaching in Ecosystem Management (a prospectus). Charles Sturt University, Albury, Australia.

Karr, J.R. (1991) Avian survival rates and the extinction process on Barro Colorado Island, Panama. *Conservation Biology*, 4 (4), 391–7.

Lawrence, W.F. (1991) Ecological correlates of extinction proneness in Australian tropical rain forest mammals. *Conservation Biology*, 5 (1), 79–89.

Luxmore, R. and Swanson, T.M (1992) Wildlife and wildland utilization and conservation, in *Economics for the Wilds* (eds T.M. Swanson and E.B. Barbier). Earthscan, London, pp. 170–94.

Panwar, H.S. (1992) *Ecodevelopment: an integrated approach to sustainable development for people and protected areas in India.* Paper presented to the Fourth World Congress on National Parks and Protected Areas, 10–21 February, 1992. Caracas, Venezuela.

Pearl, M. (1992) Conservation of Asian primates: aspects of genetics and behavioural ecology that predict vulnerability, in *Conservation Biology: The Theory and Practice of Nature Conservation Preservation and Management* (eds P.L. Fiedler and S.K. Jain). Chapman & Hall, London, pp. 297–320.

Ramsay, B.J. (1994) *Commercial Use of Wild Animals in Australia*, Department of Primary Industries and Energy, Bureau of Resource Sciences. Australian Government Publishing Service, Canberra.

Regelin, W.L. (1991) Wildlife management in Canada and the United States, in *Global Trends in Wildlife Management* (eds B. Bobek, K. Perzanowski and W. Regelin). Trans. 18th IUGB Congress, Krakow, 1987. Swiat Press, Krakow Warszawa, 1991.

Serventy, V. (1966) *A Continent in Danger*. Andre Deutsch, London.

Sidle, J.G. and Bowman, D.B. (1988) Habitat protection under the Endangered Species Act. *Conservation Biology*, 2 (1), 116–18.

Skira, I. (1992) Commercial harvesting of short-tailed shearwaters (Tasmanian mutton-birds), in *Wildlife Use and Management: Report of a Workshop for Aboriginal and Torres Strait Islander People* (eds P.D. Meek and P.H. O'Brien, Bureau of Rural Resources Report No. R/2/92). Australian Government Publishing Service, Canberra, pp. 7–18.

Soulé, M.E. (ed) (1987) *Viable Populations for Conservation*. Cambridge University Press, New York.

Swanson, T.M. (1992) The role of wildlife utilization and other policies in biodiversity conservation, in *Economics for the Wilds* (eds T.M. Swanson and E.B. Barbier). Earthscan Publications, London, pp. 65–102.

Udvardy, M.D.F. (1975) *A Classification of the Biogeographical Provinces of the World*. IUCN Occasional Paper No. 18. Morges.

UN (1992) *Long-range World Population Projections: Two Centuries of Population Growth, 1950–2150*. United Nations, Dept of International, Economic and Social Affairs, Population Division, New York.

World Commission on Environment and Development (WCED) (1987) *Our Common Future*, Report of the World Commission on Environment and Development. Oxford University Press, Oxford.

WCMC (1992) *Global Biodiversity: Status of the Earth's living resources*, compiled by World Conservation Monitoring Centre. Chapman & Hall, London.

Wilcove, D., McClellan, C. and Dobson, A. (1986) Habitat fragmentation in the temperate zone, in *Conservation Biology, the Science of Scarcity and Diversity* (ed. M.E. Soulé). Sinauer, Sunderland, Mass., pp. 237–56.

Young, M. (1992) *Sustainable Investment and Resource Use: Equity, Environmental Integrity and Economic Efficiency*. Man and the Biosphere Series, Vol. 9. UNESCO, Paris.

3

Sustainability

Sustainable use has been defined as the 'use of an organism, ecosystem or other renewable resource at a rate within its capacity for renewal' (IUCN/ UNEP/WWF 1991). The word 'use' involves everything from capture to consumption and, in animals, 'capacity for renewal' can be thought of, simplistically, as the rate at which births outstrip deaths. But because animals at birth (or hatching or metamorphosis) are not the biological equivalent of mature adults, it is more meaningful to think of 'renewal' as involving survival and growth as well as fecundity. In this sense, the rate of 'renewal' of a fish population, to the size at which the fish are harvested, would be what fisheries biologists call the **net rate of recruitment**.

3.1 GROWTH OF POPULATIONS

It is possible for resources such as nest sites or behavioural space to limit population growth while food and water are still plentiful. The limiting factors in population growth are not always obvious.

As long as no resources are limiting, healthy populations will grow as fast as the individuals of the species are able to reproduce in that particular environment – that is at their **intrinsic rate of increase**. Population growth will be exponential and the growth curve will be a steepening one like that shown for humans during this century (Figure 2.1).

If non-consumable resources (such as nesting places or display areas) become limiting, it can be expected that the population growth curve will level off abruptly when the resource limit is reached (Figure 3.1).

If a consumable resource becomes limiting, the response of the population will depend on whether the resource is able to renew itself. If all available water suddenly came to an end, for instance, a population of animals would crash to extinction; whereas if a scattered, non-renewable food supply was used up gradually, the population would decline to extinction as

Conservation and the Use of Wildlife Resources. Edited by M. Bolton.
Published in 1997 by Chapman & Hall. ISBN 0 412 71350 0.

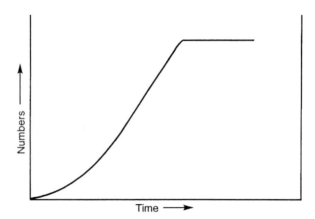

Figure 3.1 Growth of a population which is limited by a non-consumable resource such as nesting sites. The population levels off abruptly when all sites are occupied.

a decreasing number of survivors searched for food with diminishing success (Figure 3.2). But food supplies in nature are usually renewable and do not have an all-or-nothing effect on population growth. Instead, food generally influences the pattern of growth of a population of consumers according to:

- whether the consumers influence the rate at which their food is renewed
- the way in which the consumers compete for the resource when demand begins to exceed supply

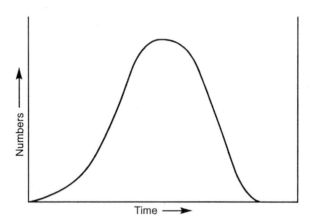

Figure 3.2 Growth and decline of a population when a consumable, but non-renewable, resource is progressively used up.

There are cases in which the rate of renewal of a food resource is independent of the number of animals using it, but all predator–prey systems show an interactive relationship between consumer and resource, and this is commonly the case in plant–herbivore systems. The way in which these interactions work can be difficult to determine and a variety of population growth patterns are possible (section 3.3.5).

At the level of individuals, consumers may respond to a resource shortage simply by showing the consequences of not getting enough of it. Like cows in a field, animals may exploit each other's food supply without showing any competitive behaviour. Eventually, as this **exploitive competition** intensifies, the animals lose condition; birth, growth and death rates are affected and the population begins to decline. This might give the food resource a chance to recover – and so support another build-up of consumers.

Competition in other species may be mediated through behavioural interference. In this scenario there are winners and losers. Territoriality is a common form of **interference competition**. Weak competitors, unable to gain or hold a territory, may be entirely deprived of an essential resource during critical times. Like red grouse (*Lagopus lagopus scoticus*) on a British moor, the individuals without territory are likely to account for most of the natural mortality (Watson and Moss 1980). Density can also affect the age groups as well as the sexes of some populations very differently, as has been found, for example, in red deer (Reynolds and Staines, Chapter 10, this volume).

The response of a population to exploitive competition for food is tightly linked to the animals' bodily needs because there is, in effect, a sharing. Interference competition is more loosely related to physical needs because individuals of some species (including red grouse) commonly defend territories which contain a surplus of the resource for which they are competing. In natural selection terms, such 'greedy' behaviour will be advantageous only as long as the reproductive benefits outweigh the costs of defending a particular territory. Although territory size is usually linked with food availability, territories have something in common with non-consumable resources, such as nesting sites, in that they separate winners from losers and so have a stabilizing effect on populations.

3.1.1   Productivity and population density

Intraspecific competition, as described above, operates in a density-dependent way in changing population size. It has the effect of suppressing a population's rate of increase at high densities, but at low densities, when competition within a species is minimal, the population is able to increase relatively rapidly. The spread of disease could also affect population size in a density-dependent manner. In theory, there must be a certain density at which additions to a population are exactly balanced by deaths so there is

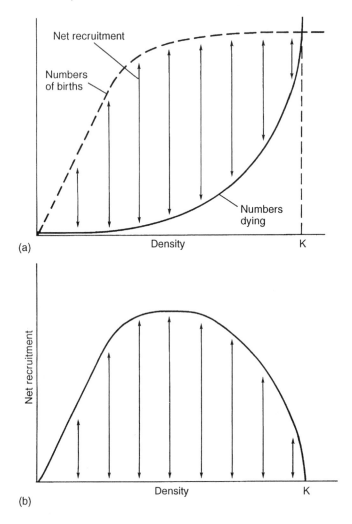

Figure 3.3   (a) Density-dependent birth and/or mortality rates will produce curves which must cross. The population density at which they cross is the carrying capacity (*K*). In nature this equilibrium point can occur over a broad range of densities and can be shifted by environmental change. (b) Net recruitment, when plotted as the difference between births and deaths at different densities, will show a humped curve.

no net change in population size. The curves for birth rate and death rate will cross on a graph at this point. It is the equilibrium density or carrying capacity – and is usually denoted by *K* (Figure 3.3a).

In natural populations *K* is not a precise or constant value. It is better thought of as the density around which populations tend to level off or fluctuate. It can be set lower or higher by density-independent factors such as the weather. Any factor that affects birth rates or death rates is a limiting factor, but only density-dependent factors have a *regulating* effect in that they tend to return a population to K from densities above or below that value.

If there are frequent disturbances, and density-dependent factors are too weak to overcome the random influences, then equilibrium may never be achieved. K becomes a moving target. Or there may be a time lag in a population's response to its own density, caused by a time lag in the response of the supporting resources. It is also possible for density-dependent factors to be strongly overcompensating so as to produce very wide fluctuations in population size. Nevertheless, it has been widely demonstrated that density-dependent effects on net recruitment do tend to keep populations within limits.

In terms of numbers rather than rates, at low densities the number of additions to a population is small; additions then increase as density rises, and decline as carrying capacity is approached, becoming negative if carrying capacity is exceeded. When net recruitment is plotted against density, therefore, the result is some form of humped or 'n'-shaped **recruitment curve** (Figure 3.3b). The recruitment curve is central to the concept of harvesting because any offtake must be replaced by recruitment if the harvest is to be sustainable. A detailed overview of intraspecific competition is provided by Begon *et al.* (1990).

3.1.2 Sustainable harvests

Sustainable harvests can be taken from a range of population densities but, in theory, the greatest harvest can be taken when net recruitment is at its maximum. This must always be at some density below K because at K deaths cancel out births and there is zero net recruitment. Once a population at K has been reduced by an initial harvest, a subsequent sustainable yield (SY) can be calculated as a fraction of the reduced size – provided that the number of animals to be taken each year is smaller than the number that was taken in the initial reduction. In Figure 3.3b the maximum sustainable yield (MSY) would be from the density at half the carrying capacity ($\frac{1}{2} K$) but in real populations the highest point on the 'n'-shaped curve may occur at a density which is much higher than 50% of carrying capacity. Values from $0.65 K$ to $0.9 K$ have been suggested for animals which breed slowly, such as marine mammals. Robinson and Redford (1991), in their model for estimating sustainable harvests from mixed-forest mammals, used a value of $0.6 K$. Figure 3.4 shows the MSY occurring at $0.65 K$.

Figure 3.4 also shows that there is only one MSY but for every level of

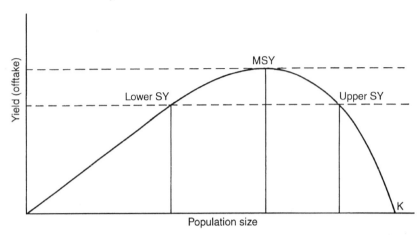

Figure 3.4 The maximum sustainable yield (MSY) can be taken when net recruitment is at its maximum; in this case it would be at 0.65K. Any sustainable yield of less than MSY can, in theory, be taken from two levels of density but it is safer to take the smaller fraction of the larger population.

SY below MSY there are two levels of density from which this SY can be harvested; one will be a large fraction of a small population and the other will be a small fraction of a large population. The latter is likely to be the more stable. Determination of population parameters and methods of calculating SY, especially for large mammals, can be found in Caughley (1977). A broader application of population dynamics to wildlife management is presented in Caughley and Sinclair (1994).

3.1.3   *K* and *r* selection

A major problem in finding high sustainable yields, even with the most cautious approach, is that populations respond to influences other than harvesting and density-dependent factors. Environmental fluctuations, being density-independent, can mask the consequences of removing a known offtake. Species for which the environment causes a high population variance need to harvested with special caution because of the risk of over-harvesting in what may happen to be a 'bad' year. The saiga antelope (*Saiga tatarica*), on the Central Asian steppes, is such a high-risk species (Milner-Gulland 1994).

It is easy to see, though, that this difficulty will apply to some types of animals much more than to others. Insects, for example, are much more vulnerable to the vagaries of the weather than are large mammals. Consequently, insects tend to have life histories which are better suited to unpredictable or ephemeral habitats; they have relatively short lives and a large allocation of energy goes into the production of numerous, rapidly maturing offspring. At the other extreme, large mammals tend to be adapted for competitive survival in relatively constant or predictably seasonal environments, having comparatively more energy invested in size, longevity, parental care, etc.

Large mammals are said to be *K*-selected species whereas species with the opposite type of life history are *r*-selected (*r* for reproduction). The distinction was made by MacArthur and Wilson (1967) between animal and plant species that were good colonizers (*r* species) and those which were better at maintaining themselves in established populations (*K* species). The concept has since been broadened into a generalization, although not all environments can be seen as obviously *r* or *K*-selecting and not all organisms conform to the scheme. The two extremes, *r* and *K*, are best regarded as the two ends of a continuum (Southwood 1976).

Many fish species appear to be *r* strategists, and density-independent mortality has been a serious obstacle to achieving sustainability in fisheries. However, there have been other reasons why fisheries have collapsed.

3.2 HARVESTING STRATEGIES

In commercial harvesting operations, whether it be fish or other animals, responsible agencies may be anxious not to over-exploit but are usually also keen not to restrict the offtake unnecessarily. It is not always the stated policy but the concept of harvesting the MSY is a popular way of thinking and it remains the basic management objective for most single-species fisheries, although 'when treated as the sole or primary objective it invariably leads to overcapacity of the fishing fleets'. (Clark 1984). As a management concept it has particularly little value when the harvested species has strong interspecific interactions (May *et al.* 1979). The term 'Optimum Sustainable Yield' sounds better and is frequently used in preference to MSY without any real distinction being made (but see section 3.5).

As a generalization, Caughley and Sinclair (1994) state that harvesting wildlife at MSY should never be contemplated. These authors favour a margin of error of at least 25% below the estimated MSY. The degree of risk of over-exploitation of a population will depend to some extent on the way in which a harvest is regulated in practice. There are a number of strategies, as follows

(a) Fixed quotas
An allowable number or tonnage is estimated and when the quota has been filled no more fishing or hunting is allowed for the season. The quota can be reviewed and changed annually. This simple method has been widely used in many different situations, from fisheries (e.g. North Sea plaice) to

kangaroos in Australia. It is particularly risky to aim for the MSY by this method because, if the MSY is overestimated by any amount, the quota will exceed recruitment rate. For common kangaroo species an annual offtake of 15–20% (based on annual surveys) has proved to be easily sustainable but fails to satisfy graziers who are concerned about the impact of kangaroos on sheep pasture. A higher quota would bring opposition from conservationists (Grigg 1995).

(b) Restricted effort
This assumes that harvesting effort can be closely matched to offtake and can be regulated. It is less risky to fix the harvesting effort (e.g. so many trawler-days, rod-days or gun-days) than to fix the quotas. If harvesting effort is expected to produce the MSY, and MSY has been overestimated (i.e. the population is smaller or less densely distributed than was thought), then fewer animals will be taken by the fixed effort.

A more or less constant level of efficiency has to be assumed in this strategy but it is possible to put an upper limit on efficiency by imposing catch limits as well as regulating the effort. In a very small scale example: abalone (*Haliotis roei*) collecting on certain beaches of Western Australia was controlled for the 1995 season by permits which authorized the holder to collect between the hours of 7.00 a.m. and 8.30 a.m. on Sundays only from 5 November to 3 December, subject to an individual daily bag limit of 20 animals.

(c) Regulated escapement
In certain circumstances, instead of fixing the harvest, it can be the size of the non-harvested portion of the population which is fixed by the management authority. A certain number of migratory ducks may be allowed to pass, for example, before shooting can begin. A constant escapement policy has been used for salmon fisheries but it requires accurate monitoring of fish stocks and can result in a negligible catch in years when recruitment has been low (Clark 1984).

(d) Dynamic pool models
In each of the preceding three strategies the exploited population was regarded, simplistically, as if it were a collection of identical animals. This ignores the fact that different sexes, age-classes and size-classes can make very different contributions to the birth rates, growth, death rates – and therefore net recruitment – of a population. In practice it is common for harvesting to concentrate on certain segments of a population (e.g. fish above a certain size, trophy stags, muttonbird chicks) rather than to take animals at random. Consequently, recruitment to the harvested categories, and recruitment to the population as a whole, are two different things.

Finally, there is an assumption that harvesting mortality directly replaces natural mortality and this may not be true. The population segment selected by the harvest is likely to be quite different from that which would have been lost by natural causes, and this could affect the survival prospects of the remainder. Ginsberg and Milner-Gulland (1994) suggest that selective cropping of male impala (*Aepyceros melampus*) could result in fewer females being inseminated – even though there may appear to be a surplus of males remaining in the herd. In addition, females which are inseminated late in the season, as a result of social disruption caused by hunting, will give birth to young which are out of synchrony with their cohort. This, the authors suggest, is likely to result in a higher mortality rate among the juveniles.

With managed ungulate herds, once the requirements are known, it is often possible to be highly selective in order to try to maintain the population structure which best serves the management objective (see Reynolds and Staines, Chapter 10, this volume). With fisheries the possibilities for selectivity are more limited and the basic population data (with few exceptions, such as migrating salmon) usually have to be obtained from sampling and estimations from catch and effort data or mark–recapture data. The age structure of a population must also be determined indirectly.

Dynamic pool strategies are reviewed by Pitcher and Hart (1982) and a classic reference work on fish population dynamics has been reprinted (Beverton and Holt 1957, reprinted in 1993). However, the spectacular failures of multi-million dollar fisheries (about a dozen in the last 50 years) have been attributable not so much to ecological complexity as to the human element in management, a point which, unavoidably, will be made more than once in this book.

3.3   RELATIONSHIPS BETWEEN SPECIES

Relationships between living things are often categorized in ecology textbooks under headings such as competition, predation, parasitism and mutualism. In nature the categories are not all clear-cut; the parasite–predator distinction is particularly blurred and in the case of mutualism, in which both partners derive benefit, varying degrees of dependency and one-sidedness have been described.

Among vertebrates very few mutually beneficial relationships have been confirmed and systematically studied, but there are innumerable examples of animals apparently gaining from the presence of others in a non-reciprocal way and the process is sometimes loosely referred to as *facilitation*. Grazing mammals respond to the alarm calls of birds; some birds recognize mammals as beaters which flush insects from the grass; many animals use the burrows and nest holes of others; scavengers feed on predators' leftovers; water holes and wallows dug by large animals are used by other creatures; forest animals use the trails created by bigger animals. This is the stuff of natural-history films but it is not often the subject of scientific study and the influence of such associations on animal numbers is generally unknown.

Competition and predation have been given much higher priority by ecologists, and under these headings it is at least possible to bring out the relevance of animal relationships to the subject of sustainable use.

3.3.1   Competition

Ecologists have given particular attention to the relationships between different species which, at first sight, seem to have similar requirements and could therefore be expected to be competing for the same resources. An assembly of different birds all searching for insects, or a variety of large mammals grazing an African plain, would seem to be competing for food just like the individuals within a single population.

It is now accepted as a general principle (the competitive exclusion principle) that if two species are in competition, they can only coexist in a stable environment as a result of not having quite the same requirements. Otherwise the stronger competitor would eventually outbreed its rival and exclude it. This is not to say that there can be no competition at all. An animal's preferred 'place' in the ecological scheme of things, its **fundamental niche**, may not be fully realized in competition with other species. Its **realized niche** could be somewhat limited as a result of some **niche overlap** with other species. Evidence for this has been obtained experimentally for a range of species by removing one (or more) of the competing types from a site and monitoring the population responses of the species remaining. An increase in abundance as a response would indicate that the species had been in competition to some extent. In this way, Hairston (1980) was able to demonstrate that two species of salamander, which coexist on the Southern Appalachian Mountains of the USA, could each do better (in population terms) in the absence of the other.

Expressed in everyday terms, different animals with overlapping requirements are obviously able to live together successfully – but they may be cramping each other's style.

3.3.2   Close niches and facilitation

There are instances of facilitation between animals in very close niches. One example has been described by Bell (1971) for grazing animals in the Serengeti. In certain parts of the Serengeti, during the wet season, short grasses predominate on the tops of the ridges when on lower ground coarse tussock grasses have reached their full height. The resident grazing animals feed on the ridge tops until the start of the dry season, when they begin to move down into the longer grasses. Buffaloes (*Syncerus caffer*) and zebras (*Equus burchelli*) are among the first in this grazing succession for they can eat the tallest, toughest stems. This makes the basal leaves more accessible to wildebeest (*Connochaetes taurinus*) and topi (*Damaliscus korrigum*), which in turn leave a short sward for the little Thomson's gazelle (*Gazella thomsoni*).

3.3.3 Niche separation

As with intraspecific competition, the interaction between individuals of different species can be exploitive or it can take the form of interference. It can also be highly asymmetric in that the consequences are not the same for both species.

Precisely what separates two similar niches can be hard to discover because the niche includes all aspects of the environment that can affect an animal's (or plant's) chances of passing on its genes to the next generation. In animals the separation may consist in something as subtle as a preference for eating a different stage in plant growth or having a slightly different breeding season. Or it could be a combination of such factors. If **niche differentiation** cannot be revealed, its existence may just have to be assumed in order to explain how two or more demonstrably different species are able to coexist when they seem to be in direct competition. Sometimes there may be no niche overlap, and no interspecific competition, even though niches appear to be similar. Coexistence of species in similar niches can be the evolutionary outcome of competition in the distant past.

However, not even the principle of competitive exclusion can be applied in every situation. When an ecological gap is suddenly created by events such as landslides or fire, it is possible for colonizing species to coexist for a time with species that would normally exclude them. In the definition of the competitive exclusion principle, the term 'stable environment' leaves open the question of time-scale. In all living relationships, environmental changes can be overriding at any time.

3.3.4 Introduced species

It is to be expected that animals that have evolved together in a natural community will have arrived at an optimal degree of niche separation as a result of the whole system having been weighed in the balance for so long. Animals which have not evolved together may be occupying niches which are not so efficiently packed. In an experimental area feral buffaloes (*Bubalus bubalis*) were eliminated from 300 km^2 of Australia's monsoonal north. There was a significant increase in the number of feral pigs (*Sus scrofa*) in response. The pig population was thought to have been limited by interference competition from buffaloes which compacted the ground at a time when young pigs were rooting for food (Corbett 1995). It is hard to believe that pigs, given evolutionary time, could not have evolved a more favourable foraging and time-sharing relationship with buffaloes. Buffaloes were introduced to Australia about 160 years ago, when pigs were probably already feral in the area.

3.3.5   Predators and prey

Even at the level of one predatory species and its prey, numerical relationships are not always what they seem and cause and effect can be hard to determine. If prey are living at high density, then predation may thin out their numbers and reduce intraspecific competition. In response, the prey population may compensate for the losses by increasing more rapidly. But among vertebrates it is often the individuals least likely to survive and reproduce which are killed by predators – in which case there may not be much impact on the prey population dynamics. And such impact as there is will be inversely density-dependent because removing unfit individuals from high density prey populations would reduce food competition and improve conditions for the remaining breeding stock. There are several theoretical possibilities.

(a) *Limitation of prey by predators*
If prey species increase in number after predation has been eliminated or reduced, as in removal experiments, the indication is that the predators had been limiting their prey populations. In some cases predators do reduce the harvest of prey species available to humans; cormorants on an English reservoir, for example, were found to be depressing anglers' catch rates of trout (Roberts 1991). In this case both the anglers and the cormorants were competing for well-grown fish. In other cases, humans and other predators could be selecting different segments of a prey population.

(b) *Regulation of prey by predators*
Whether predators *regulate* prey populations in a density-dependent manner, around an equilibrium density, is another question. When predators are allowed to return to an area from which they have been removed, one might expect the prey species to return to a lower density. But this does not always happen. In Australia, fox removal-and-return experiments indicated that rabbits were able to break away from regulation by foxes (Pech *et al.* 1992).

There is some evidence for **multiple equilibria** between predators and prey with unstable thresholds in between. When prey species, especially those with a high intrinsic rate of increase such as rabbits, are allowed to increase above a critical density, their populations may be beyond predatory control because predatory mammals and birds are usually limited not only by their own intrinsic rates of increase but by interference competition that results in spacing. This pattern has been revealed in studies of North American moose (*Alces alces*) and wolves (*Canis lupus*). An upper limit to wolf density (58.7 animals per 1000 km^2) was probably set by spacing behaviour. The studies suggest that wolves cannot limit moose populations if moose density is higher than 0.65 animals per km^2. Moose densities of up to 2.5 animals per km^2 have been recorded, at which point food would be limiting (Messier 1995).

Conversely, a prey population which has been reduced below a critical density may be unable to recover if predators maintain a high searching intensity – especially if prey refuges or escape strategies have been adversely affected by habitat change.

(c) Response of predators to prey density

The last points raise the topic of a predator's feeding behaviour in relation to prey density. This is known as the **functional response**, as distinct from the **numerical response** of populations. Clearly, it is impossible for predators to maintain a high population density if they reduce prey numbers to the point where the predators cannot find enough to eat. Much will depend, therefore, on whether the predator has an alternative prey species and at what point predators switch attention to prey B when prey A is at low densities. In Australia, for example, dingoes are able to maintain relatively constant populations by switching seasonally between a variety of prey species (Corbett 1995).

The question whether selection of prey B takes the predatory pressure off prey A, when the latter is in decline, could be a question of survival or local extinction for the declining prey. Predation models for primary and secondary prey species have been been reviewed by Pech *et al.* (1995).

(d) Predator–prey cycles

In the northern hemisphere predators and herbivores can show cyclic changes in abundance. It appears as though predators reduce prey numbers to the point where predator density must decline, thereby allowing the prey density to increase again, supporting another predator build-up, and so on. But interlinked cycles are not necessarily generated by the predator–prey relationship; instead they may be reflecting a density-dependent relationship between the herbivore and its food supply, with the predator numbers merely tracking those of the prey. Over the years both models have been applied to the 10-year cycles shown by populations of the American snowshoe hare (*Lepus americanus*) and the lynx (*Lynx canadensis*). These exceptional cycles, which occur in synchrony over vast areas, have been familiar to generations of fur trappers, and have become something of an ecological classic in that they have attracted study and speculation for most of the present century. A recent study has demonstrated a mechanism which combines the effects of both food and predation on the snowshoe hares.

Foraging by hares is necessarily a compromise between finding enough to eat and avoiding predators. There is evidence that, during the low phase of the hare cycle, hares try to avoid predation to such an extent that they do not obtain an adequate diet even while sufficient food exists. The consequent loss of condition is another factor in the population decline (Hik 1995).

(e) Summary

The changes in abundance of predators and prey can show a variety of patterns, with or without human interventions. Populations are not regulated by any single mechanism but rather respond to a web of constantly shifting interactions. Simple explanations are mainly derived from mathematical models or laboratory experiments, and these can only be tentatively applied to the complexities of observed patterns in the fields and forests; indeed, existing field data can be used to support different interpretations (e.g. Boutin 1995).

3.4   THE PROBLEMS OF MULTISPECIES MANAGEMENT

Not surprisingly, in a number of fisheries the decline of one species has been followed by an increase in another and the causal relationships have rarely been clear. Nobody can say to what extent the changes have been due to compensatory shifts in population densities, changes in patterns of harvesting or even coincidental environmental effects (May *et al.* 1979).

Tropical fisheries can produce dozens of species in a single trawl and as the multiple relationships are likely to remain unknown it might seem sensible to manage such fisheries for a multispecies yield rather than to target particular taxa at the risk of disrupting the entire system. But as Clark (1984) has pointed out, raw fish can differ in commercial value by two orders of magnitude so that the notion of a multispecies MSY is commercially meaningless. Clark goes on to suggest that perhaps: 'the central unresolved problem for multispecies fishery management is the problem of finding a saleable, operational catch-phrase to replace MSY, or if that proves impossible, an understandable series of basic management principles'. This problem, which has troubled fishery biologists for a long time, is discussed by May *et al.* (1979). It is not a problem unique to fisheries; rather it is a monumental example of the specialization dilemma discussed in Chapter 2: how to crop wildlife economically – and in the proportions delivered by nature.

3.5   THE SUSTAINABILITY TRIANGLE

Anthony Charles (1994) has reviewed the changing paradigms in fishery management and noted that by the 1950s the biologists' perspective, or

'Conservation Paradigm' had come to be generally accepted. In this view, ecological concerns took precedence over human objectives in fishery management. Protecting fish stocks was recognized as a necessity although MSY was still the obvious goal to pursue.

During the 1950s this view was challenged by a 'Rationalization Paradigm' with an emphasis on economic efficiency. Management strategies were aimed at producing the Maximum Economic Yield (MEY). Clark (1973) pointed out that if carried to the extreme, this view could even rationalize extinctions, for if returns on invested money are higher than biological growth rates for an exploited species, then achieving the MEY might involve converting the species to cash and reinvesting the capital in the city.

Charles (1994) describes a third paradigm which emphasizes concern for the social fabric of fishing communities. From this perspective, biological sustainability is essential but the way in which a sustainable yield is distributed is more important than its magnitude, no matter whether it is measured in fish or money. This gives some meaning to the notion of 'Optimal Sustainable Yield' for it allows for a range of human concerns in addition to conservation goals. On this basis Charles expresses the idea of sustainability, as the simultaneous pursuit of four components:

- ecological sustainability
- socio-economic sustainability, measured at the level of individuals and aggregated across the resource system
- community sustainability, recognizing that a community is more than a collection of individuals
- institutional sustainability, centred on the financial, administrative and organizational arrangements by which the other three components are maintained

In this construction the first three components can be viewed as the points of a sustainability triangle, held together by institutional sustainability (Figure 3.5). The concept has practical application in that it provides a framework for sustainability assessment. In trying to achieve sustainability in any wildlife utilization scheme, not just fishing, there will be crucial questions to be asked, and criteria to be met under the headings of each of the four components. This procedure will be taken further in Chapter 15 but at this point the recent collapse of a major fishery will serve as an example to bring together the points already made.

3.5.1 Collapse of a fishery

In 1968 the catch of cod (*Gadus morhua*) from the Grand Banks of Newfoundland was 810 000 tonnes. By 1977, despite what were believed to be safe quotas, the catch has fallen to 150 000 tonnes. The Canadian government blamed quota cheats and foreign fishing vessels. Accordingly, a 200-

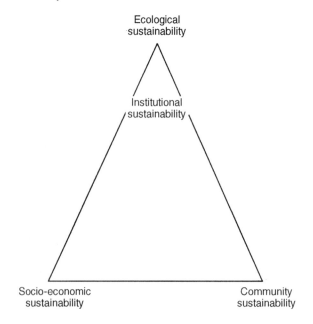

Figure 3.5   The sustainability triangle (redrawn from Charles 1994).

mile limit was enforced to keep out foreign vessels, and total allowable catches (TACs) were set at 16% of estimated populations in order to allow stocks to recover. Instead, catches continued to decline while political tensions over the issue increased. In 1992 the Canadian Department of Fisheries and Oceans (DFO) set a TAC of 120 000 tonnes but, later in the year, banned fishing altogether. By that time the fishing communities themselves had agreed to a moratorium which, in 1993, was extended indefinitely. Fishing was no longer providing an income for families which had known no other way of life.

In explanation, it has been reported that the DFO research vessels made annual random trawls in order to estimate fish population size whereas commercial trawlers do not trawl at random. Commercial fishermen do their best to locate the fish stocks and concentrate their fishing effort in the more productive areas. Not surprisingly, there were wide discrepancies between DFO and commercial catches per unit effort. In the economic and socio-political circumstances which prevailed, compromise estimates were adopted and it has been suggested that this was one factor in the collapse of the fishery (Mackenzie 1995).

At the ecological level it is still not understood why there should have been such a drastic failure in recruitment. A codfish produces millions of eggs so it was thought reasonable to assume that a tiny difference in survival rate could easily compensate for a large reduction in the number of egg-producing adults. However, hardly anything is known about events such as migration, mating and spawning of fish at ultra-low densities in the wild. Observations of spawning cod and haddock in aquaria suggest that 'mating in these species may be a less prosaic affair than is usually assumed – and its success perhaps at risk if densities are too low' (Beverton 1984). Perhaps predation is involved; or perhaps, as in some other social animals, these shoaling fish breed less successfully at very low densities – an effect described by W.C. Allee in 1931. In solitary species an Allee effect might result simply from the difficulty of locating a mate but in social animals it can have a behavioural basis. It has been shown, for instance, that ovarian development in the female ring dove (*Streptopelia risoria*) is accelerated by the noises of the dove colony (Lott *et al.* 1967). Whatever the ecological explanation, it will be only one critical factor at one point of the sustainability triangle.

3.6   SUSTAINABLE DEVELOPMENT

'Sustainable development' has become a catch-phrase of the environmental movement, but, like many successful slogans, it means different things to different people. Sometimes 'sustainable development' is taken to be more or less synonymous with 'sustainable use' (e.g. Wildes 1995) but the word 'development' implies a change in status of the user and it is this which causes all the difficulty in attempts at definition.

The World Commission on Environment and Development (WCED 1987) characterized sustainable development in terms of pursuing those paths of social, economic and political progress that meet 'the needs of the present without compromising the ability of future generations to meet their own needs'. This is frequently quoted as a definition but it is not very helpful to conservation because nobody knows how far the abilities of future generations will be changed through technological advance.

A more biological definition is offered by IUCN/UNEP/WWF (1991) in which sustainable development means 'improving the quality of human life while living within the carrying capacity of supporting ecosystems'. But, again, this focuses attention on the extent to which different ecosystems are supportive of humanity.

Perhaps the term 'sustainable development' became popular because it seems less of a contradiction in terms than 'sustainable growth'. But in its open-endedness, development is not far removed from growth and in global terms the two are inextricably linked through human population expansion.

Unlike 'sustainable use', which can be defined in terms of particular resources and rates of renewal, 'sustainable development' is a phrase that seeks to reconcile socio-cultural and socio-economic ambitions with the constraints of ecology. The emphasis is squarely upon people and those species and processes of nature which are necessary for human support. But this is at the heart of the biodiversity debate: what is necessary for human support? There is no agreement about the importance of individual species and ecosystems. Nobody can say precisely how much biodiversity we require to provide the material basis of human life, let alone predict all the non-essential resource values of the future. And when ethical, aesthetic and other personal values are taken into account it is not difficult to see why 'sustainable development' has never been defined in a way that is generally acceptable. For a discussion of the conservation–development dilemma see Robinson (1993) and Holdgate and Munro (1993).

REFERENCES

Allee, W.C. (1931) *Animal Aggregations: A Study in General Sociology*. University of Chicago Press, Chicago.

Begon, M., Harper, J.L. and Townsend, C.R. (1990) (2nd edition) *Ecology: Individuals, Populations and Communities*. Blackwell Science, Cambridge, Mass.

Bell, R.H.V. (1971) A grazing ecosystem in the Serengeti. *Scientific American*, 225 (1), 86–93.

Beverton, R.J.H. (group rapporteur) (1984) Dynamics of single species, in *Exploitation of Marine Communities* (ed. R.M. May), Report of the Dahlem Workshop on Exploitation of Marine Communities, 1–6 April 1984, Berlin. Springer Verlag, Berlin, pp. 13–58.

Beverton, R. and Holt, S. (1957) *On the Dynamics of Exploited Fish Populations*. Reprint with amendments, 1993. Chapman & Hall, London.

Boutin, S. (1995) Testing predator–prey theory by studying fluctuating populations of small mammals. *Wildlife Research* 22 (1), 89–100.

Caughley, G. (1977) *Analysis of Vertebrate Populations*. John Wiley, Chichester.

Caughley, G. and Sinclair, A.R.E. (1994) *Wildlife Ecology and Management*. Blackwell, Cambridge, Mass.

Charles, A.T. (1994) Towards sustainability: the fishery experience. *Ecological Economics*, 11, 201–11.

Clark, C.W. (1973) The economics of overexploitation. *Science*, 181, 630–4.

Clark, C.W. (1984) Strategies for multispecies management: objectives and constraints, in *Exploitation of Marine Communities* (ed. R.M. May), Report of the Dahlem Workshop on Exploitation of Marine Communities, 1–6 April 1984, Berlin. Springer Verlag, Berlin, pp. 303–12.

Corbett, L. (1995) Does dingo predation or buffalo competition regulate feral pig populations in the Australian wet–dry tropics? An experimental study. *Wildlife Research*, 22 (1), 65–74.

Ginsberg, J.R. and Milner-Gulland, E.J. (1994) Sex-biased harvesting and population dynamics in ungulates: implications for conservation and sustainable use. *Conservation Biology*, 8 (1), 157–66.

Grigg, G. (1995) Kangaroo harvesting for conservation of rangelands, kangaroos . . . and graziers, in *Conservation Through Sustainable Use of Wildlife* (eds G.C. Grigg, P.T. Hale and D. Lunny). Centre for Conservation Biology, University of

Queensland, Brisbane, pp. 161–5.

Hairston, N.G. (1980) The experimental test of an analysis of field distributions: competition in terrestrial salamanders. *Ecology*, **61**, 817–26.

Hik, D.S. (1995) Does risk of predation influence population dynamics? Evidence from the cyclic decline of snowshoe hares. *Wildlife Research*, **22** (1), 115–29.

Holdgate, M. and Munro, D.A. (1993) Limits to caring; a response. *Conservation Biology*, **7** (4), 938–40.

IUCN/UNEP/WWF (1991) *Caring for the Earth: A strategy for sustainable living.* Gland, Switzerland.

Lott, D., Scholz, S.D. and Lehrman, D.S. (1967) Exteroceptive stimulation of the reproductive system of the female ring dove (*Streptopelia risoria*) by the mate and by the colony milieu. *Animal Behaviour*, **15** (4), 433–7.

MacArthur, R.H. and Wilson, E.O. (1967) *The Theory of Island Biogeography.* Princeton University Press, Princeton, NY.

Mackenzie, D. (1995) The cod that disappeared. *New Scientist*, **147**, 24–9.

May, R.M., Beddington, J.R., Clark, C.W., Holt, S.J. and Laws, R.M (1979) Management of multispecies fisheries. *Science*, **205**, 267–77.

Messier, F. (1995) Trophic interactions in two northern wolf–ungulate systems. *Wildlife Research*, **22** (1), 131–46.

Milner-Gulland, E.J. (1994) Sustainable management of the saiga antelope. *Oryx*, **28** (4), 257–62.

Pech, R.P., Sinclair, A.R.E., Newsome, A.E. and Catling, P.C. (1992) Limits to predator regulation of rabbits in Australia: evidence from predator-removal experiments. *Oecologia*, **89**, 102–12.

Pech, R.P., Sinclair, A.R.E. and Newsome, A.E. (1995) Predation models for primary and secondary prey species. *Wildlife Research*, **22** (1), 55–64.

Pitcher, T.J. and Hart, P.J.B. (1982) *Fisheries Ecology.* Croom Helm, London.

Roberts, K.A. (1991) Field monitoring: confessions of an addict, in *Monitoring For Conservation and Ecology* (ed. F.B. Goldsmith). Chapman & Hall, London, pp. 179–211.

Robinson, J.G. (1993) The limits to caring: sustainable living and the loss of biodiversity. *Conservation Biology* **7** (1), 20–8.

Robinson, J.G. and Redford, K.H. (1991) Sustainable harvest of Neotropical forest animals, in *Neotropical Wildlife Use and Conservation* (eds J.G. Robinson and K.H. Redford). University of Chicago Press, Chicago, pp. 415–29.

Southwood, T.R.E. (1976) Bionomic strategies, in *Theoretical Ecology: Principles and Applications* (ed. R.M. May). Blackwell Scientific Publications, Oxford, pp. 26–48.

Watson, A. and Moss, R. (1980) Advances in our understanding of the population dynamics of red grouse from a recent fluctuation in numbers. *Ardea*, **68**, 103–11.

WCED (1987) *Our Common Future*, Report of the World Commission on Environment and Development. Oxford University Press, Oxford.

Wildes, F.T. (1995) Recent themes in conservation philosophy and policy in the United States. *Environmental Conservation*, **22** (2), 143–50.

4

Conservation, controversies and concerns

Around the world people express very different concerns about wildlife, and the general public may have quite different perceptions from those of professional conservationists. In a survey among the American and Japanese public, for instance, only 30% and 18%, respectively, regarded habitat destruction by humans as a major threat to wild species. Nor are attitudes and opinions entirely predictable from measures of education and national development. The Japanese, by and large, were found to harbour a 'mastery of nature' outlook which westerners would probably now consider to be naive and outdated. In Japan this attitude was not strongly influenced by level of education, whereas in the USA college-educated people expressed greater ecological and ethical concern for animals than did those of lower educational attainment (Kellert 1991).

The distinction between ecology and ethics can present a conservation problem in itself because those expressing ecological concerns can find themselves in conflict with those who are more worried about the ethics of exploitation. A wide range of views is to be expected from the variety of relationships which now exist between humans and other animals, but conservationists need to find common ground. This chapter examines some of the issues surrounding field sports, the use of protected areas, and the rights and welfare of animals.

4.1 FIELD SPORTS

In terms of the number of participants, field sports make up the largest commercial sector in wildlife utilization. In 1990 in the United Kingdom some 3.9 million people went fishing and there were about 0.8 million hunters (i.e. game shooters and those who 'hunt' in the English sense of the

Conservation and the Use of Wildlife Resources. Edited by M. Bolton.
Published in 1997 by Chapman & Hall. ISBN 0 412 71350 0.

Table 4.1 Some national comparisons of sport hunting

Country	Number of hunters (thousands)	% Hunters in total population	Shooting licence test[*]
USA	14 100	5.5	0-0-0
France	1 650	2.9	1
Spain	1 450	3.7	0/1
Italy	950	1.7	1
United Kingdom	800	1.5	0–0–0
Germany	326	0.4	1–2–3
Sweden	320	3.7	1–2–3
Finland	296	5.8	1–2–3
Greece	295	2.9	1
Portugal	243	2.4	1–2
Denmark	177	3.4	1–2–3
Ireland	120	3.3	0–0–0
New Zealand	117	3.4	Under review
Austria	94	1.4	1–2–3
Netherlands	33.5	0.2	1–2–3
Belgium	29	0.3	1–2
Luxembourg	2.3	0.6	1–2–3
Australia		2.5–5	Under review

[*]0 None; 1 written; 2 practical; 3 shooting test.
Sources: US Government 1993 for USA. FACE 1995 for Continental Europe and Ireland. Cobham Resource Consultants 1992 for the UK. Nugent and Fraser 1993 for New Zealand. Figures not available for Australia and range given represents 15–30% of estimated number of gun owners.

word – with hounds, but excluding clay pigeon shooters). Direct expenditure on British field sports exceeded £1.4 billion (*c.* US$2.12 billion) (Cobham Resource Consultants 1992).

In the European Union some 6.8 million hunters spend an estimated £8.23 billion (*c.* US$12.4 billion) on their sport each year and maintain about 100 000 jobs in the process (FACE 1995). In 1991 over 14 million Americans went hunting and 35.6 million Americans (some of them also hunters) went fishing. Between them these people spent US$41 billion on their sport during the year (US Government 1993). Table 4.1 shows numbers of hunters by country.

Field sports range from specialized and extremely expensive safaris in the more remote parts of the world to casual fishing or rough shooting in the local countryside. There may also be an element of sport in subsistence hunting; a pig hunt in New Guinea is not just a matter of collecting pork; it is obviously an exciting event for the men and dogs involved and a successful hunt will be relived around the evening fire. King (1994) mentions that hunting by villagers in west Cameroon is also considered to be both an economic and a recreational pursuit.

4.1.1 Game animals

The animals pursued as 'game' around the world range from elephants to small passerine birds. A species may, at different times or in different countries, have legal status as game, protected animal or vermin. Caughley and Sinclair (1994) make the point that traditional game animals are characterized by having a high intrinsic rate of increase and strongly density-dependent population growth.

4.1.2 Sport and conservation

Unlike commercial hunters, recreational hunters are not economically motivated and sport hunting has an 'enviable record of conserving hunted stocks. Instances of gross overexploitation are rare but not unknown' (Caughley and Sinclair 1994). There are other reasons for the special relationship between field sports and conservation.

(a) Easy and effective control
It is in the nature of all sports that participants wish to feel proud and are generally ready to adopt restrictions and rules about what is acceptable. Field sports are unique in wildlife utilization in that they actually promote inefficient methods of taking. The angler tries to tempt trout with a fly on a fine line when it would be much easier to clean out the pool with a net across the stream. There may also be an incidental policing value in the presence of legal shooting parties. Eltringham (1994) mentions that poaching has increased in countries where legal hunting has been stopped.

(b) Conservation inputs: revenue
Relatively common and unspectacular animals can be made to generate revenue far in excess of their carcase value because hunters and anglers are paying for a recreational experience, not just for fish or meat. Days when nothing is bagged can still be enjoyed. In New Zealand, for example, recreational deer hunters spend more than NZ$ 300 (c. £138), on average, to kill a deer with a venison value of about NZ$ 125 (c. £57) (Nugent and Fraser 1993). The deer stalker in Scotland must pay a £200–250 fee to shoot a quality red deer stag (Reynolds and Staines, Chapter 10, this volume).

Revenue collected as rents, rates, membership fees, licence fees etc. can be used for conservation purposes and this is commonly the case in western countries. In the USA, revenue collected by the states from hunting licences must, by law, be spent on wildlife management.

(c) Conservation inputs: labour and pressure groups

Hunters and anglers are often prepared to work as unpaid labour to improve field conditions for their sport. This may not always be beneficial for species other than the preferred game but it can be so, especially where the purpose is to ameliorate the impacts of intensive agriculture or to restore health to wetlands and waterways. In Tasmania, property-based game management plans were introduced in 1993 and can involve deer hunters in habitat management work as part payment for hunting access to a property (Murphy 1995).

Some field sports associations, such as the National Wildlife Federation in America, are politically powerful enough, along with other conservation organizations, to influence governments to the benefit of wildlife habitat.

4.1.3 Ecological impacts of field sports

One of the key assumptions in support of wildlife utilization as conservation strategy is that it can provide incentives for protecting wildlife habitat. In western countries some of the clearest support for this assumption is derived from the influence of field sports on agricultural practices. In the interests of game preservation, farmers may leave some 'rough' even in the most intensively cultivated landscapes (see Aebischer, Chapter 8, this volume).

Unfortunately field sports are not free from the ecological problem of specialization (section 2.3.3), and management can be weighted so heavily in favour of the preferred animals that competitors are not tolerated even when there is no evidence that they are detrimental to the sport. In Europe generations of gamekeepers have displayed on their gibbets the corpses of mammals and birds killed on the assumption the sport would be better without them; an assumption that was sometimes quite unfounded and is still common. The director of the Main Inspectorate of Game Management in Poland, for example, stated that: 'Animals enjoying the status of protected species constitute another group of vermin . . . and will mention among them goshawks and both species of buzzards. . . . Within the range of their occurrence one should effectively eliminate foxes, racoon-dogs, badgers, pine martens and polecats' (Sikorski 1991).

Whereas field sports can offer unique opportunities for conservation, the downside associated with specialization is, of course, common to agriculture, forestry, and all other situations in which animals come to be seen as pests.

4.1.4   Systems of control

Ownership of the resource exerts a fundamental influence on wildlife utilization, including sport hunting. In North America, in the 50 years between 1850 and the turn of the century, wildlife exploitation exemplified the tragedy of the commons in that huntable species were decimated in a free-for-all of killing and wilderness taming. The estimated 60 million bison (*Bison bison*) that existed in 1860 were represented by about 150 survivors in 1889 (Robinson and Bolen 1984, cited in Regelin 1991). These animals were slaughtered for their hides and to reduce the food supply of the Indians. Other game animals were hunted to the point of scarcity for meat and sport. The passenger pigeon (*Ectopistes migratorius*) became extinct. Predators, notably the wolf (*Canis lupus*), cougar (*Felis concolor*) and grizzly bear (*Ursus arctos*) were extirpated from most of continental US and southern Canada in the interest of livestock protection (Regelin 1991).

In most cases, American game species have now recovered their numbers. White-tailed deer (*Odocoileus virginianus*) are actually more numerous and occupy a larger range than at any time in recorded history; more than three million a year are harvested by sport hunters (Regelin 1991). In the opinion of Geist (1988, 1994) the recovery of North American game animals can be attributed to a few core policies of wildlife management, especially the first three of the following:

- public ownership of native wildlife
- egalitarian allocation of hunting rights by law at a cost affordable to most people
- prohibition of trade in dead wildlife and animal parts (excluding furs)
- prohibition of frivolous and wasteful killing
- involvement of wildlife professionals in management

These points are worth considering in an international context.

(a) Ownership and hunting rights

Indisputably, public ownership is preferable to open-access commons from a conservation viewpoint but opinions differ about the relative merits of public and private ownership in regard to wildlife management; many different systems having been made to work. Most western European hunting grounds are privately owned but some large areas are administered by the government, or are under communal ownership, or may be reserved for members of specified associations – such as the French Associations Communales de la Chasse Agrées. In the UK hunting rights belong to the landowner, however small the property, whereas in Belgium and the Netherlands hunting rights are subject to a minimum area (C. King, pers. comm.).

Under the revier system, practised in Germany and Austria, hunting blocks (reviers) may be managed by the owners or leased to individuals or associations who then become legally responsible for all hunting activities on the block – including damage compensation. Those holding the responsibility (landowner or otherwise) must be qualified by examination to hold a revier-operator's licence and are answerable to official hunting inspectors. An obvious advantage of this system is that game management is conducted to 'official' standards with minimal cost to the government (Bubenik 1989).

Nordic countries established private hunting rights centuries ago but land can be leased for elk (syn. moose *Alces alces*) hunting only if the owner has obtained government approval to declare the land an elk hunting area. Populations of elk are now probably larger than ever previously recorded (Haagenrud *et al.* 1987).

In a view from Zimbabwe, Child (1995) has argued that the success of the North American system has depended heavily on the philanthropy of private and native landowners who have borne the cost of wildlife production but have been unable to capture much of its great value. This appears to be a view shared by an increasing number of American and Canadian landowners, who are pressing for more private control over hunting. In some states there have been moves in this direction, and in other states, although private landowners have no wildlife management authority as such, they can and do charge for the privilege to hunt or fish and may exercise control through the laws of trespass. In Colorado, an increasing number of cattle ranchers are managing deer and providing accommodation and other facilities for fee-paying hunters. Other private landowners have entered into franchise agreements with their state wildlife authority to provide recreational hunting under management objectives jointly developed by the state and private managers (Davis 1995).

Geist (1988) argues that increasing private control will result in higher costs, lower participation rates, diminished public support for hunting and, as a result, adverse consequences for conservation. There are obviously a great many facets to this argument and there are opposing views even among North American wildlife managers.

Unquestionably, public opinion is an extremely important factor, but, like the wildlife itself, it is something that usually has to be managed within the social, economic and political frameworks which happen to prevail in any manager's part of the world. The private versus public ownership argument is, therefore, rather academic for wildlife managers in the field.

(b) Sale of meat and other parts

Sport hunting, western-style, requires reasonable wealth and leisure and the existence of game species. Where these requirements coexist, or can be brought together through organized hunting trips, it may be possible to emulate the USA and Canada by building a successful industry based on sustainable utilization without legal sale of dead wildlife. The main advantage of keeping wildlife products out of the market-place is that the incentive for commercial production remains limited, and illegal trade (poaching) is easier to control. Geist (1988) compares the heavy cost of gamekeeping in Germany with the relatively light cost of game protection in Alberta, Canada, and attributes the difference largely to the absence of an American and Canadian market in game meat.

But the sale of game meat is an established and important element of game management in Europe and many other countries, and a ban on wildlife meat and other products in countries with little or no sport hunting potential could preclude any conservation efforts based on wildlife utilization. Again, it is not possible to draw general conclusions from the particularly fortunate circumstances of the USA and Canada.

(c) Wasteful killing

Geist (1988) points out that because recreational hunting arouses more public disapproval than 'the old American idea of hunting for food', the use of the term 'sport' is a threat to conservation. In North America and Canada an animal killed by a hunter must not be abandoned under penalty of law, so in a practical sense the killing has not been frivolous or wasteful. However, it is difficult to believe that those who disapprove of hunting could be convinced that the average western hunter *needs* to kill for food so it seems pointless to try to maintain that for most western hunters there is not a recreational element in their pot-hunting.

(d) Involvement of professionals

It goes without saying that any system of management is likely to benefit from the honest application of good science. In North America the opportunities have been realized and a high standard of scientific and managerial expertise is publicly funded for wildlife management. But there are plenty of other successful arrangements and outside North America publicly funded experts work in cooperation with the private sector, as in the management of Scotland's red deer (Reynolds and Staines, Chapter 10, this volume).

Some of the most pervasive problems of wildlife management, however, are not technical at all. In the words of Chester Phelps (1984): 'Perhaps the unstable foundation that underlies all of the current complexities [of wildlife management] is that in the United States, fish and wildlife are no longer born free. From the moment of parturition, wild creatures are subject to the rules, regulations, policies, interests and whims of man . . . the problems are largely people problems – not fish and wildlife problems'.

To illustrate his point, Phelps recounts that in the state of Virginia it took more than 25 years to convince a generation of sportsmen and policy makers that there was no value in rearing and releasing bobwhite quail (*Colinus virginianus*) in habitat where the quail were already at carrying capacity.

4.1.5   Pests and resources

It is not unusual for game or other commercial species to be actual or potential pests. A policy of pest control by shooting or trapping is really the same thing as harvesting, and unless the annual offtakes exceed MSY the control measures will have the same effect as sustainable hunting. Moreover, common pests and traditional game species tend to have similar population characteristics and show strong recovery from moderately low densities. For this reason, shooting and trapping may not be a cost-effective way of meeting control objectives.

It might seem sensible to encourage recreational hunting of potential pests in order to reduce the cost of control operations. In the UK, during the 1950s and early 1960s, the government subsidized shotgun cartridges to be used against woodpigeons (*Columba palumbus*). However, shooting up to 40% of the pigeons in winter did not reduce the numbers occurring in study areas the following spring. Shooting was only cropping a proportion that was destined to die anyway from winter food shortage (Murton *et al.* 1974).

Perhaps a really massive shooting effort would have depressed woodpigeon populations but that possibility introduces another reason why it is difficult to manage a species simultaneously as a pest and as quarry for field sports. If an animal exists at a density high enough to satisfy recreational hunters, then it is likely to be too numerous to please those who regard it as a pest. In New Zealand this dilemma centres on introduced deer, especially the red deer (*Cervus elaphus scoticus*).

There are about a quarter of a million wild deer in New Zealand, supporting an annual harvest of some 70 000 animals. Nearly two-thirds of these are taken by 34 000 recreational hunters. Browsing by deer, even in very low densities, suppresses the regeneration of preferred plant species – especially sub-canopy hardwoods – in New Zealand's forests. Those who most value the native ecosystems would therefore like to see deer exterminated or kept down to the lowest possible densities. This would offer very poor hunting prospects and it is not an objective that would ever have the support of the hunters. It has been suggested that the conflict between hunters and conservationists could be reconciled by dividing the country into management units, ranked according to conservation values, so that the objective of conservation and hunting could be pursued separately in the most appropriate areas of forest (Nugent and Fraser 1993). The principle of zonation is used extensively in this way to cater for incompatible activities within protected areas.

4.2   WILDLIFE USE AND PROTECTED AREAS

Protected areas (PAs) are, or should be, established for good reasons and if the reasons for protecting a particular area is plainly stated then management has a clear objective. It has become fashionable to say that 'parks are for people' but the other 97% of the Earth's surface is also for people so the important thing is to be clear about why parks are special.

Controversies about resident people in PAs often seem to be based on philosophical or political positions as if there were no scientific criteria to be considered. This can be a consequence of having no clearly stated purpose for protection. Where the goal of management has been agreed and adopted, then political, philosophical and ethical decisions about the status of the area should already have been made and the level of scientific management should have been reached.

This is not to suggest that scientifically based management decisions will always be right or easy but when the goal is clear there is at least a chance of recognizing and correctly weighing the pros and cons of management alternatives.

4.2.1 Harvesting versus other activities

In PAs of many parts of the world it is common to find that livestock grazing and even crop cultivation are tolerated by the protection authorities but hunting and trapping are regarded as more serious offences. Ecologically, depending on all the circumstances, the controlled harvesting of wildlife could be far less damaging than the other activities mentioned. It is possible to drag hundreds of thousands of muttonbird chicks from their burrows every year with no detriment to the population, whereas trampling by domestic animals could destroy the colony (section 2.4).

If a policy of minimal interference has been adopted for an area, such as would be the case for a Strict Nature Reserve, then harvesting and all other exploitive activities cannot and should not be contemplated. In areas with lower overall protection status, the ecological significance of different human activities can only be assessed in relation to the aims of management of the particular site.

4.2.2 The disturbance factor

Because of the complexity of animal relationships it may not be possible to avoid ecological collateral damage even when a species is harvested sustainably. This may be unimportant in an environment which is already modified for agriculture or forestry, but in a PA the ecological side effects might be less acceptable than the direct loss through harvesting. Damage to riverside vegetation by anglers may be of more concern than the harvest of fish. Fishermen near crocodile nesting grounds in Ethiopia inadvertently destroyed crocodile nests by causing parent crocodiles to leave the site, which in turn allowed varanid lizards to dig up the eggs and eat them.

The disturbance factor is sometimes overlooked or too readily discounted in PAs (see also Chapter 14). It has been suggested, for example, that domestic livestock in India's Gir forest, home of Asia's only remaining lions, do not compete with wild ungulates because the latter are mainly browsers and the domestic animals are mainly grazers (Berwick 1976, cited in Raval 1991). Other possibilities are:

- Interference competition from livestock influences what wild ungulates are able to eat. This effect would be exacerbated by the presence of herd boys and dogs.
- Domestic herds modify the vegetation and partly determine what food is available for deer and other herbivores at all times.

Exclosures and stock-removal experiments would throw more light on controversies of this sort.

4.2.3   The question of scale

The size of a PA can determine what sort of management goals are possible. At the scale of a single hectare the fall of a large tree is a major disruption, but over a few square kilometres of forest it is possible to see tree deaths as part of a shifting pattern within a higher order of relative stability. And yet a lightning fire could destroy hundreds of square kilometres so if processes on that scale are to be seen as part of nature's shifting pattern, within a greater mosaic of relative stability, then we would need to look at thousands of square kilometres. In managing a protected area there will be a choice between trying to preserve the existing pattern, and trying to let natural processes run their course. Peak population densities of large mammals in PAs can devastate areas of natural vegetation. Should the population then be culled to keep the status quo? The decision can arouse strong emotions, especially when elephants are involved, but if we want to let nature run its course and still preserve the grand scheme of things, then we need to have a lot of nature.

The most common size for a protected area on a world-wide basis is only 10–30 km^2 (Johnston 1992).

4.2.4   Area size, animal size and harvest sustainability

In Chapter 1 the notion of animal sources and sinks was introduced in connection with subsistence hunting. The concept is especially relevant to PAs or other discrete habitats where sustainable use is contemplated. The scale of operations must be in proportion to the areas and animal densities involved. In the Arabuko-Sokoke forest of northern Kenya, for example, elephant shrews were trapped in the forest periphery. It was estimated that this reduced the peripheral shrew density by 40% but removed only 4% of the total population. Shrews could be replaced from the forest interior. But primates and other large mammals had no refuge and no security because they were actively hunted throughout the forest. For the most part, the forest is less than 15 km across so the interior is within a day's hunting range of the periphery, which is where the people live (Fitzgibbon *et al.* 1995). For the primates the forest is all sink and no secure source.

4.2.5 Buffer zones

At the scale of landscape, animals are able to distribute themselves in response to the balance of their pressures and needs but the same pressures and needs often force animals beyond the boundaries of protected areas.

Ideally, large mammals leaving a protected area would find themselves in a buffer zone. The buffer zone would be a less attractive habitat to the animals because a lower level of protection would operate and there would be some human activity such as collection of forest produce. The buffer zone would not be cultivated. Animal species at high densities inside the reserve could use the buffer as a dispersal area, in which case cropping from that zone need have no detrimental effects. That is the theory.

Ideal situations are rarely encountered. The reserves of Africa and Asia are becoming surrounded and increasingly penetrated by cultivation and livestock. Wild herbivores are often attracted to food crops growing outside a protected area so that culling those animals represents a drain on the core population. Fences are an expensive prevention measure. Large predators, forced out by spacing behaviour, find themselves immediately among livestock.

Domestic animals make up 35% of the kills of the Gir forest lions and 28 people were killed by them between 1978 and 1991 (Saberwal *et al.* 1994). It is usual to find that 'problem' predators are sub-adults trying to disperse. If there is nowhere else to put them, these 'surplus' Gir lions might have to be shot. That raises the theoretical question of whether safari hunters should be allowed to pay handsomely for the privilege (revenue to be used for conservation of course), or whether that would be so controversial as to do more harm than good. It is an academic point because safari hunting is not legal in India.

4.2.6 People and protected areas in Africa

In a questionnaire survey among people living adjacent to six of Tanzania's PAs, 71% of respondents reported problems with wildlife. The familiar pattern that emerged was that at high human densities large mammals were scarce and smaller pests such as birds, monkeys, pigs and rodents were proving difficult to control (Newmark *et al.* 1994). In the Arabuko-Sokoke Forest of Kenya the baboons (*Papio cynocephalus*) and Syke's monkeys (*Cercopithecus mitis*) are hunted with dogs because they are a threat to crops outside the forest; the local people do not eat the meat (Fitzgibbon *et al.* 1995). This makes it more difficult to turn these pests into a harvested resource.

There is no guarantee that benefits from wildlife utilization will be acceptable compensation for the perceived disadvantages of living adjacent to any particular PA. Botswana still has comparatively low human population densities, and villagers in a study area near the Chobe National Park (depending on their circumstances) were issued with special game licences which permitted them to kill some large mammals at very little cost. Holders of these licences took about 295 kg of meat annually. The Park and associated wildlife-based enterprises also provide a large proportion of the local employment. None the less, it was found that local attitudes to wildlife were largely negative, with most respondents recording complaints about crop damage and risk of personal injury from wild animals. Perceptions of benefits were generally low. The financial return, per person, from animals shot was calculated to be over nine times as much as that perceived by the villagers. Similarly, returns from hunting were perceived by the villagers to be more important than those from employment, when employment was calculated to be worth 20 times more to the villagers than was personal hunting (Parry and Campbell 1992). Perhaps if the local people had been actively involved in management they would have been more appreciative of the benefits. The Wildlife Conservation and National Parks Act, passed in 1992, does enable communities living within non-reserve wildlife areas to play a stronger role in wildlife management (Steiner and Rihoy 1995).

The community component of management has been specifically targeted in some African initiatives. The Communal Area Management Programme for Indigenous Resources (CAMPFIRE) in Zimbabwe has been particularly well publicized (e.g. Barbier 1992; Child 1995; Child 1996) and is evidently making good progress, both in strengthening local participation in wildlife management and in returning a higher proportion of wildlife benefits to local people. Some 69 000 households (about 550 000 people) were involved in the scheme in 1993. More than 90% of the benefits derive from safari hunting (64% of this from elephant hunting) with eco-tourism beginning to emerge in a few prime sites (Child 1995; Steiner and Rihoy 1995). Scale is considered to be critical; management units must consist of fewer than 200 households if they are to be successful, and the units of management, benefit and authority should be one and the same. The state retains only a regulatory role (Child 1996).

Even under CAMPFIRE, however, there has been dissatisfaction with the proportions of the returns which are retained by some district and ward councils (Steiner and Rihoy 1995). As Child (1995) has remarked, it is relatively easy to earn money from wildlife but much more difficult to get the proceeds allocated correctly. It does make sense, when dividends are insufficient for distribution to households, to spend the money on small capital projects for community benefit, and the best solution for the present appears to be a policy of permitting community members to select between personal and community needs on the basis of a majority vote (Child 1995).

4.2.7   Wildlife use in protected areas: some conclusions

Harvesting wildlife is not always in conflict with the aims of protected area management. Depending on the objectives of management, agricultural activities can be much more harmful than controlled harvesting. However, objectives and circumstances are site-specific and it is not intended here to promote wildlife harvesting inside areas which can be afforded protection without such interference. If any generalization is to be made, it can be said that there is a need for less, not more, disturbance in protected areas.

Outside PAs it is to be hoped that the conservation benefits of utilization *will* help to counter the further conversion of natural habitats to other forms of use. In countries with similar wildlife assets to those of Botswana and Zimbabwe, this could well apply to land adjacent to protected areas. However, where the policy is not appropriate, attempts to involve local residents in a utilization scheme could actually strengthen local hostility towards a PA by appearing to confirm what the locals already suspected: that immediate, tangible benefits do not outweigh the immediate, tangible costs.

4.3   ANIMAL RIGHTS AND ANIMAL WELFARE

This very broad area of concern focuses mainly upon animals as individuals. Conservation, in contrast, is concerned with numbers. Conservation efforts must be applied at the level of animal populations with the survival of species in mind. Sometimes, when numbers are low, the interests of the individual and the species are much the same thing. Some animals have become so rare that virtually every individual is important to the survival of the species. At the other extreme, if weaker individuals did not die in large numbers, entire populations would die from mass starvation.

As we have seen, however, conservation is not pure biology and a great many people, including this writer, have a concerned interest in animals, both as species and as individuals. The two concerns, welfare and conservation, are sometimes in conflict: when pests such as rabbits are being controlled by a viral disease, for instance, there may be indisputable suffering while the interests of conservation are being served. Failing to recognize, and separate, the different aims and concerns leads to muddled thinking. This section can do nothing more than frame the issues.

4.3.1   Animal rights

The case for animal rights requires an acceptance that rights are intrinsic and not merely considerations which humans choose to extend to other species. Some writers in the 'deep ecology' movement would extend rights to the whole of nature but most advocates of animal rights would probably argue only on behalf of those animals which are capable of knowing and feeling – the sentient creatures.

There is a continuity through higher animals and unless we believe in supernatural interventions there is remarkably little difference between humans and the nearest non-humans. In the 8 million years (it could be less) since the human and 'other ape' lines diverged, we have shed very little of our common ancestry. With only about 2% difference between the genes of chimpanzees and humans, the chimps are closer to humans than they are to gorillas – which evidently branched off rather earlier. Such is the affinity between humans and chimps that, in the laboratory, it is technically possible to implant a fertilized chimp egg into an emptied human blastocyst so that a chimp could develop inside a human surrogate mother (as a chimp, not a 'humanzee') with a compatible human placenta (Cherfas 1986). No doubt, as Diamond (1991) asserts, a zoologist from space, reclassifying the superfamily Hominoidea on the basis of DNA, would not put *Homo sapiens* in a separate family.

The most conspicuous differences between chimps and people are the trappings of a few thousand years of civilization. Presumably, this is why it is easier to think of Palaeolithic man as being closer to our ancestors, even though he was genetically the same animal as ourselves. As Dawkins (1986) has pointed out, it is convenient for our legal and moral systems that the embarrassing intermediates, between humans and our common ancestors with the apes, are all dead.

The case for animal rights cannot easily be dismissed on philosophical grounds. If rights were to be extended beyond the species barrier, however, people would have to agree on just where to draw the line; on what levels of animal awareness or social complexity should be taken into account. The problem is explored in some detail by Rosemary Rodd (1990) who believes that: 'accepting that at least some animals possess at least some definite rights to moral attention involves fewer violations of our intuitive perceptions than an obstinate insistence that only humans can be right-holders'. This may strike a chord with many readers, but nowhere in the world is it likely to improve the status of wild animals in any significant way.

4.3.2 Animal welfare

In what Broom and Johnson (1993) refer to as an 'informed and compassionate society' it should be possible to accept some responsibilities for the welfare of other species, whether or not non-humans have intrinsic rights. This stewardship ethic, with ourselves as moral agents, does not seem to stretch the existing framework of moral values, although Christianity has not been helpful in this respect. According to Singer (1990) the New Testament is completely lacking in any injunction against cruelty to animals, or any recommendation to consider animals' interests. This would not have surprised René Descartes, the French philosopher (1596–1650), because in his understanding only humans have souls; consciousness is evidence of a soul; therefore animals do not have consciousness and can feel no pain. In the Cartesian view, which had a strong influence on western society, animals were just automata which could give a 'realistic illusion of agony'. This particular reference to Descartes can be found under 'cruelty to animals' in the *Encyclopaedia Britannica*.

The world's first anti-cruelty law was not passed until 1822, when Richard Martin persuaded the British House of Commons to pass what came to be known as the Martin Act. Two years later the world's first animal welfare society was formed which, upon receiving royal patronage in 1840, became the RSPCA. It is now accepted, on scientific evidence, that animals can have complex subjective feelings and that they can suffer. See, for example, Dawkins (1980).

The main thrust of anti-cruelty legislation has been directed towards domestic animals; a bill to protect wild mammals from gratuitous cruelty in the United Kingdom was not passed by the British House of Commons until 1996.

It is surprisingly difficult to define 'welfare' but Broom and Johnson (1993) offer the following: 'The state of an individual as regards its attempts to cope with its environment'. In this definition welfare is a characteristic of an animal, not something given to it. There is a continuum of welfare states and these should be measurable in a way that is independent of moral considerations. Welfare is not quite the same thing as health or fitness because fit and healthy animals can still be afraid or in pain. Neither can welfare be entirely equated with pain and suffering because there are circumstances in which animals with very poor welfare could not be conscious of pain. It is also true that poor welfare can result from frustration and under-stimulation as well as over-stimulation. The indicators of welfare states are shown in Table 4.2.

4.3.3 Animal welfare and conservation

As discussed in Chapter 3, animals in the wild commonly die as a result of predation or starvation rather than old age. Animals which die peacefully and comfortably in the wild must be the exception. It seems reasonable to say that, as moral agents, we should not increase the sum of suffering unnecessarily and should make every effort to provide for the welfare of animals which are under our direct care. It is not realistic, however, to think of conserving animal populations without killing individual animals. As Eltringham (1994) has remarked, 'a distaste for slaughter should not be disguised as an objection to cropping on conservation grounds'.

It is becoming increasingly difficult to protect wildlife from exploitation or displacement. Directly and indirectly, humans will continue to be responsible on a vast scale for the deaths of wild animals. But where killing and capturing can be made subject to some degree of control, there can be opportunities to ensure that the operations are done humanely. At this

Table 4.2 Indicators of animal welfare states (source: Brcom and Johnson 1993)

Welfare indicator	Very good	⟶	Very poor
Adrenal cortex activity	Occasional bouts of adrenal cortex activity	Frequent adrenal activity. Higher synthetic enzyme levels	Pathological consequences of adrenal activity, eventually associated with reduced possibility of adrenal activity
Stereotyped behaviour	Occasional stereotypy caused by minor frustration	Stereotypies for 5% of active time	Stereotypies for 40% of active time
Growth Reproduction Life expectancy	Normal growth and reproduction	Impaired growth or reproduction	Impaired growth or reproduction and reduced life expectancy
Suffering as result of injury	No injury	Injury – asleep or narcotized	Injury – awake or suffering
Immune system function and disease condition	Normal immune system functioning	Substantial immunosuppression	Substantial immunosuppression plus severe disease condition

point, conservation and animal welfare interests can be brought together. When human involvement with animals is formalized, it becomes possible to introduce and enforce regulations. In the field, trapping and slaughter methods can be regulated. Shooters can be directed to use appropriate calibre weapons and ammunition. Licensed shooters can be tested for marksmanship and knowledge. Shot size can be specified in order to achieve cleaner kills. These things may not happen, but at least they become possible with legal, structured forms of wildlife use.

In Australia, the RSPCA found that commercial kangaroo shooters had the best record for clean killing, followed by non-commercial but legal shooters. The highest incidence of suffering was caused by illegal shooters (Australian Government 1988). Despite this fact, the voice of protest against kangaroo shooting is directed mainly against commercial activities and this raises the final point for this section: much of the public controversy about wildlife utilization appears to be centred not on either conservation or welfare issues at all, but on moral judgements. Morality is a human construct whereas welfare is in the experience of individual animals. It is the *manner* in which an animal lives or dies that is important for its welfare; not the *motives* of the killer or keeper, nor whether he comes from a village in New Guinea or a penthouse in New York.

The muddle of conservation, welfare and moral concerns is exemplified by a comment from the Director of the Wildlife Preservation Society of Queensland: 'Most members of conservation groups find shooting animals for the fun of it a pretty disgusting idea and not one that fits well with efforts to improve the relationship between humans and their environment' (Jeffreys 1995). The conservation groups referred to were Australian and in the same article Jeffreys reported that most of the 15 or so major state and national groups had no overall policy on the sustainable use of wildlife for conservation.

ACKNOWLEDGEMENTS

Thanks to Charles King in Brussels, Stella Rees of the British Association for Shooting and Conservation, and Peter Allen of the *Australian Shooters Journal* for providing me with useful statistics for this chapter. Thanks, too, to Trish Larner for library help.

REFERENCES

Australian Government (1988) *Kangaroos*. Report by Senate Select Committee on Animal Welfare. Parliamentary Paper 109 of 1988. Australian Government Publishing Service, Canberra.
Barbier, E.B. (1992) Community-based development in Africa, in *Economics for the Wilds* (eds T.M. Swanson and E.B. Barbier). Earthscan, London, pp. 103–35.

Berwick, S. (1976) The Gir Forest: an endangered ecosystem. *American Scientist*, **64**, 28–40.

Broom, D.M. and Johnson, K.G. (1993) *Stress and Animal Welfare*. Chapman & Hall, London.

Bubenik, A.B. (1989) Sport hunting in continental Europe, in *Wildlife Production Systems; Economic Utilization of Wild Ungulates* (eds R.J. Hudson, K.R. Drew and L.M. Baskin). Cambridge University Press, Cambridge, pp. 115–33.

Caughley, G. and Sinclair, A.R.E. (1994) *Wildlife Ecology and Management*. Blackwell, Cambridge, Mass.

Cherfas, J. (1986) Chimps in the laboratory: an endangered species? *New Scientist*, 27 March 1986, 37–41.

Child, B. (1996) CAMPFIRE in Zimbabwe, in *Assessing the Sustainability of Uses of Wild Species – Case Studies and Initial Assessment Procedure* (eds R. Prescott-Allen and C. Prescott-Allen). IUCN, Gland, Switzerland, and Cambridge, UK, pp. 59–78.

Child, G. (1995) *Wildlife and People: the Zimbabwean Success*. Wisdom Foundation, Harare.

Cobham Resource Consultants (1992) *Countryside Sports, their Economic and Conservation Significance*. Standing Conference on Countryside Sports, Reading, UK.

Davis, R.K. (1995) Using markets to achieve wildlife conservation, in *Conservation Through Sustainable Use of Wildlife* (eds G.C. Grigg, P.T. Hale and D. Lunny). Centre for Conservation Biology, University of Queensland, Brisbane.

Dawkins, M. (1980) *Animal Suffering: The Science of Animal Welfare*. Chapman & Hall, London.

Dawkins, R. (1986) *The Blind Watchmaker*. Longman, Harlow, UK.

Diamond, J. (1991) *The Rise and Fall of the Third Chimpanzee*. Radius, London.

Eltringham, S.K. (1994) Can wildlife pay its way? *Oryx*, **28** (3), 163–8.

FACE (1995) *Handbook of Hunting in Europe*, Fédération des Associations de Chasseurs de l'UE, Brussels.

Fitzgibbon, C.D., Mogaka, H. and Fanshawe, H. (1995) Subsistence hunting in Arabuko-Sokoke Forest, Kenya, and its effects on mammal populations. *Conservation Biology*, **9** (5), 1116–26.

Geist, V. (1988) How markets in wildlife meat and parts, and the sale of hunting privileges, jeopardize wildlife conservation. *Conservation Biology*, **2** (1), 15–26.

Geist, V. (1994) Wildlife conservation as wealth. *Nature*, **268**, 7 April 1994, 491–2.

Haagenrud, H., Morow, K., Nygrén, K. and Stålfelt, F. (1987) Management of moose in Nordic countries, in *Swedish Wildlife Research; Proc. 2nd Internat. Moose Symposium*. Stockholm, pp. 635–42.

Jeffreys, A. (1995) NGOs and sustainable use, in *Conservation through Sustainable Use of Wildlife* (eds G.C. Grigg, P.T. Hale and D. Lunny). Centre for Conservation Biology, University of Queensland, Brisbane, pp. 29–34.

Johnston, S. (1992) Protected areas, in *Global Biodiversity: Status of the Earth's living resources* (compiled by Wildlife Conservation Monitoring Centre). Chapman & Hall, London, pp. 447–478.

Kellert, S.R. (1991) Japanese perceptions of wildlife. *Conservation Biology*, **5** (3), 297–308.

King, S. (1994) Utilisation of wildlife in Bakossiland, west Cameroon, with particular reference to primates. *Traffic Bulletin*, **14** (2), 63–73.

Murphy, B.P. (1995) Management of wild fallow deer in Tasmania: a sustainable approach, in *Conservation Through Sustainable Use of Wildlife* (eds G.C. Grigg, P.T. Hale and D. Lunny). Centre for Conservation Biology, University of Queensland, Birsbane, pp. 307–11.

Murton, R.K., Westwood, N.J. and Isaacson A. (1974) A study of woodpigeon shooting: the exploitation of a natural animal population. *Journal of Applied Ecology*, **11**, 61–84.

Newmark, W.D., Manyanza, D.N., Gamassa, D.M. and Sariko, H.I. (1994) The conflict between wildlife and local people living adjacent to protected areas in Tanzania: human density as a predictor. *Conservation Biology*, **8** (1), 249–55.

Nugent, G. and Fraser, K.W. (1993) Pests or valued resources? Conflicts in management of deer. *New Zealand Journal of Zoology*, **20**, 361–6.

Parry, D. and Campbell B. (1992) Attitudes of rural communities to animal wildlife and its utilization in Chobe Enclave and Mababe Depression, Botswana. *Environmental Conservation*, **19** (3), 245–52.

Phelps, C.F. (1984) Bio-politics and the mature professional, in *Natural Resource Administration: Introducing a New Methodology for Management* (eds C.W. Churchman, A.M. Rosenthal and S.H. Smith). Westview Press, Boulder, Colorado, pp. 147–54.

Raval, S.R. (1991) The Gir National Park and the Maldharis: Beyond "setting aside", in *Resident Peoples and National Parks: Social dilemmas and strategies in international conservation* (eds P.C. West and S.R. Brechin). University of Arizona Press, Tucson, pp. 68–86.

Regelin, W. (1991) Wildlife Management in Canada and the United States, in *Global Trends in Wildlife Management* (eds B. Bobek, K. Perzanowski and W.L. Regelin). Trans. 18th IUGB Congress, Krakow, 1987. Swiat Press, Krakow-Warszawa, Poland, pp. 55–64.

Robinson, W.L. and Bolen, E.G. (1984) *Wildlife Ecology and Management*. Macmillan, NY.

Rodd, R. (1990) *Biology, Ethics and Animals*. Clarendon Press, Oxford.

Saberwal, V.K., Gibbs, J.P., Chellam, R. and Johnsingh, A.J.T. (1994) Lion–human conflict in the Gir Forest, India. *Conservation Biology*, **8** (2), 501–7.

Sikorski, J. (1991) Big game management in Poland – economical aspects and perspectives, in *Global Trends in Wildlife Management* (eds B. Bobek, K. Perzanowski, and W.L. Regelin). Trans. 18th IUGB Congress, Krakow, 1987. Swiat Press, Krakow-Warszawa, Poland.

Singer, P. (1990) *Animal Liberation* (2nd edition). Jonathon Cape, London.

Steiner, A. and Rihoy, E. (1995) The commons without the tragedy. Strategies for community based natural resources management in southern Africa. Background paper to Proceedings of the Regional Natural Resources Management Programme Annual Conference. SADC Wildlife Techical Coordination Unit, Lilongwe, Malawi, pp. 9–43.

US Government (1993) *National Survey of Fishing, Hunting and Wildlife-Associated Recreation, 1991*, US Dept of the Interior, Fish and Wildlife Service, and Dept of Commerce, Bureau of the Census, Washington DC.

Part Two

Case Studies and Categories,
A Selection

5

Giant clams: mariculture for sustainable exploitation

John S. Lucas

5.1 INTRODUCTION

5.1.1 Exploitation of giant clams

Giant clams are among the bottom-inhabiting wildlife in shallow habitats of coral reefs. Because of their shallow distribution, they are readily located and collected, and they have been a traditional and reliable food source throughout their geographic ranges. The shells of giant clams are used for containers (e.g. pig troughs), tools and ornaments. Giant clams are collected opportunistically during general food searches in the intertidal zone and by free-diving. Some Pacific communities even accumulate giant clams in 'shell gardens' as a food source when unfavourable weather prevents other fishing activities.

However, in recent decades the exploitation of giant clams has been at unsustainable levels throughout much of their geographic ranges. The causes of this increased exploitation will be discussed later. One important result of concern about severe over-exploitation and the status of giant clam populations has been a surge of research interest. The research has been directed in particular at the mariculture of giant clams as a measure for sustainable exploitation and for reef restocking of the wildlife resource.

5.1.2 Giant clam species

Either eight or nine living species of giant clams are recognized at present (Table 5.1). They occur in the tropical and subtropical latitudes of the Indo-Pacific region and the focus of their distributions is the Indo-Malay region, where most species occur. One species, *Tridacna rosewateri*, is known only from the Saya de Malha Bank, western Indian Ocean, whereas

Conservation and the Use of Wildlife Resources. Edited by M. Bolton.
Published in 1997 by Chapman & Hall. ISBN 0 412 71350 0.

Table 5.1 Living species of giant clam

Species, author (common name)	Max. shell length (cm)	Distribution (status)
Tridacna gigas Linnè, 1758 (Giant clam)	100+	SE Asia to Micronesia (heavily fished except off Great Barrier Reef; some extinctions)
T. derasa Röding, 1798 (Smooth giant clam)	50+	SE Asia to western Micronesia and Tonga (heavily fished; some extinctions)
T. tevoroa* Lucas, Ledua and Braley, 1991 (Evil giant clam)	50+	Tonga and Fiji (rare)
T. squamosa Lamarck, 1819 (Fluted giant clam)	40	E. Africa to eastern Melanesia (heavily fished)
T. maxima Röding, 1798 (Rugose giant clam)	30	E. Africa to eastern Polynesia, except Hawaii (heavily fished to abundant)
T. rosewateri[†] Sirenko and Scarlato, 1991 (Rosewater's giant clam)	?	One locality, W. Indian Ocean (only known from its type locality)
T. crocea Lamarck, 1819 (Burrowing giant clam)	15	Japan, SE Asia and Micronesia (heavily fished to abundant)
Hippopus hippopus Linnè, 1758 (Horse's hoof giant clam)	50+	SE Asia to eastern Melanesia (heavily fished)
H. porcellanus Rosewater, 1982 (China giant clam)	50+	Western Pacific (heavily fished)

*T. tevoroa, Lucas, Ledua and Braley, 1991, is probably a junior synonym of T. mbalavuana Ladd.

[†]T. rosewateri is only known from the type specimens, which are shells, and its status is not clear.

T. maxima is distributed from the east African coast across the Indian and Pacific Oceans as far as the eastern islands of French Polynesia. No species occur in the Atlantic Ocean, in the Caribbean Sea nor on the west coast of the Americas.

Only one giant clam species, *T. gigas*, is truly gigantic (Figure 5.1). The largest shell length reported is about 137 cm and the heaviest specimen about 500 kg (Lucas 1994). Four other giant clam species may grow to

Figure 5.1   A large *Tridacna gigas* at Arlington Reef, Great Barrier Reef. It would weigh several hundred kilogrammes.

more than 50 cm in shell length (Table 5.1), but they only weigh up to about 15 kg whole weight; a magnitude less in weight than *T. gigas*. Thus, *T. gigas* is the only species that truly warrants the name 'giant' and it has the distinction of apparently being the largest bivalved animal in evolutionary history (Yonge 1975). This gives a special reason for conserving this species, if one should be needed.

5.1.3   Biology of giant clams

Giant clams are markedly different from other bivalved molluscs (oysters, mussels, etc.). Their mantle lobes (which line the shells' interior) project beyond the shells as a large, fleshy organ (Figure 5.1). This exposed mantle is packed with millions of symbiotic microalgae per gramme of tissue. The microalgae, known as zooxanthellae, transfer organic products from their photosynthesis to the clam host. The clam reciprocates with inorganic nutrients and by exposing the zooxanthellae to sunlight. (This need for sunlight explains why giant clams are limited to shallow depths.) Much of the giant clam's nutrition is obtained from these symbiotic algae. However, giant clams also filter-feed with their gills, like those of bivalved molluscs.

Giant clams have life cycles that are like other bivalved molluscs (Figure 5.2), with several unusual features. Firstly, fecundity of the larger species is enormous. Fecundity ranges from millions of eggs per spawning in small *T. crocea* to hundreds of millions of eggs in *T. gigas*. In fact, several spawnings of *T. gigas* of approximately 1000 million eggs have been recorded (Lucas 1994), putting this species near, if not at, the pinnacle of animal fecundity. Secondly, they are protandric hermaphrodites. They mature as males after approximately 2–6 years (taking longest in *T. gigas*) and as females a year or more later. Mature clams have hermaphrodite gonads consisting of ovaries and testes mixed throughout. Yet the clam sheds spermatozoa and then eggs as distinct events, separated by an hour or more, during spawning. Thirdly, their eggs are without zooxanthellae, but they ingest zooxanthellae as larvae and early juveniles, and commence symbiosis during their early juvenile development.

Giant clam eggs are planktonic and soon hatch as trochophore larvae, which develop into bivalved veliger larvae (Figure 5.2). These larvae swim and filter-feed on microalgae (phytoplankton). The veligers grow, gain a

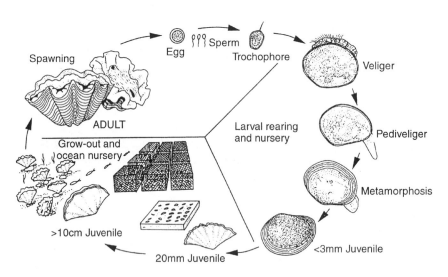

Figure 5.2 Stages in the life cycle and mariculture of giant clams (from Calumpong 1992; reproduced with permission).

prominent foot and are known as pediveliger larvae. These settle out of the plankton on to selected reef surfaces and metamorphose into juvenile clams. The juvenile clam has a foot and may creep around a little, but it is very small (200 μm shell length) and where it settles is essentially its habitat for life.

Initial growth is slow in absolute terms and juvenile clams reach about 5 cm shell length in a year. Thereafter, there is rapid growth for at least several years until the growth rate begins to taper off. As a result, the size/age growth curve is sigmoidal, which is like the pattern in many other benthic invertebrates. However, while growth rate declines in older clams, growth never ceases. Annual growth increments are evident as annual pairs of bands in cross-sections of the shell (Lucas 1994).

Contrary to popular belief, large giant clams are not hundreds of years old. The oldest one that has been reliably aged is a huge specimen of *T. gigas* estimated at 63 years (Lucas 1994).

5.2 UTILIZATION OF GIANT CLAMS

5.2.1 Causes of over-exploitation

As indicated in the Introduction, giant clams have been utilized throughout their geographic ranges by traditional societies. This has not necessarily been excessive exploitation and the smaller species can still be found at high densities in parts of their range. However, it appears that even in pre-historic times giant clam species were exploited to extinction in areas of large human populations, such as southern Japan. This is indicated by the presence of shells of species where they are no longer found.

The widespread over-exploitation of giant clam populations in recent decades is a new phenomenon caused by a combination of factors (Box 5.1). The relative importance of these five factors will vary in different regions and has not been quantified.

As an example of trade in giant clam meat in a traditional society, Fiji Fisheries' statistics show that the sale of giant clam meat (mantle and adductor muscle) on municipal markets, wholesale and retail outlets, and exports totalled 277.5 tonnes over a 10-year period to 1988.

Poaching by foreign fishing vessels has been the particular prerogative of Taiwanese clam boats. These boats have ranged through the Indo-Malay region, the Great Barrier Reef and Pacific islands. They have most affected populations of the larger giant clam species, which they target. Divers locate clams and cut out the big muscle (the adductor muscle) that pulls the two shells together. This muscle accounts for about 10% of the clam's flesh; the remaining flesh is left to rot and be eaten by the local reef carnivores.

Giant clam muscle is highly prized in Chinese cuisine and very valuable, hence the motivation for illegal fishing. Landings of giant clam adductor

---

Box 5.1   Causes of recent declines in wild stocks of giant clams

1.   Increased human populations

2.   Improved technology available to subsistence fishing people (e.g. power boats and diving gear increasing effective fishing effort)

3.   Expanded inter-island and international trade in giant clam meat, including poaching by foreign fishing vessels

4.   Expanded international trade in shell specimens and artefacts, and aquarium specimens

5.   Reef habitat degradation (e.g. pollution, siltation, and mechanical damage)

---

muscle in Taiwan in the 1960s and 1970s have been variously estimated as between 100 and 400 tonnes per annum (Carleton 1984; Dawson and Philipson 1989). The lowest estimate, 100 tonnes per year, corresponds to 300 000–450 000 clams per year, indicating a harvest of millions of clams over the decades of intense international poaching in the 1960s and 1970s.

Even the Great Barrier Reef received attention from the giant clam poachers until prosecutions and government pressure were effective. It has been estimated that approximately 2 million *T. derasa* and *T. gigas* were poached by foreign vessels between 1969 and 1976 (Pearson 1977).

In the Philippines, the main economic importance of giant clams is their shells. The Philippines has been the source of most of the international trade in giant clam shells and shell artefacts. All species are used, but the *Hippopus* species and *T. squamosa* are most popular because of their ornateness. Despite the International Union for Conservation of Nature (IUCN) ban on international trade in giant clam products, giant clam shells are exported to Japan, Australia, Europe, USA, etc. (Juinio *et al.* 1987). Recorded exports over an 8-year period were 2225 tonnes, corresponding to about 2.28 million shell pairs.

5.2.2   Giant clam fisheries in global context

The consistent pattern of commercial fisheries on giant clams is of over-fishing, with giant clam stocks declining precipitously and sources becoming increasingly scarce or remote. The giant clam fisheries, however, are only minor components of regional mollusc production. For example, annual production through fisheries and mariculture of bivalve molluscs for the adductor muscle market in Japan exceeds 200 000 tonnes per annum (Dawson and Philipson 1989). The hundreds of tonnes per annum of giant clam meat, at the peak of international poaching, are trivial compared to other mollusc fisheries and mariculture industries. Thus, it is not that giant clams have been overwhelmingly fished compared with other molluscs. The problems with commercial fisheries of giant clams, especially the larger species, are that giant clams usually occur at low densities, have low and erratic recruitment rates, and take years to reach the size at which they are being harvested. Thus, there is little support for establishment of sustainable commercial fisheries of giant clams based purely on wild stocks.

5.3 CONSERVATION OF GIANT CLAMS

5.3.1 Measures

Five measures for assisting the recovery of overfished giant clam populations have been implemented or proposed (Box 5.2).

Measure 1 was implemented with the listing of giant clams in The IUCN Invertebrate Red Data Book of threatened invertebrates (Wells *et al.* 1983), followed by the listing of all species on Appendix II of CITES. Yet, there is still major international trade in giant clam meat, shells and live specimens for aquaria, with material derived from wild stocks. In particular, Taiwan is a recipient of meat, and the Philippines and Indonesia are sources of shells and live specimens.

5.3.2 Results of conservation measures

The results of measures 2–4, where they have been implemented, are difficult to judge. Establishing these measures in a region of over-exploited populations probably means that levels of recruitment will be low and population recovery slow. Establishing a local breeding population (measure 4) may result in recruitment to reef areas a substantial distance away, due

---

Box 5.2   Measures for assisting the recovery of wild stocks of giant clams

1. A ban on international trade in giant clam products

2. Marine reserves with temporary or permanent bans on fishing giant clams

3. Limitations on fishing effort and harvestable size

4. Aggregating the remaining clams so that their reproduction will be facilitated by proximity for fertilization of gametes when they breed

5. Restocking with cultured giant clams

to dispersal of larvae. There needs to be careful monitoring of population densities over a number of years and this has rarely been undertaken. Some population recovery was achieved in a small marine reserve on Ishigaki Island, southern Japan (Murakoshi 1987), but this was only over many years.

Measure 5 will be considered in the next section. This was a major factor in the early research into the mariculture of giant clams.

5.4   MARICULTURE FOR CONSERVATION

5.4.1   Mariculture of giant clams

As noted earlier, there is little support from previous experience for sustainable commercial fisheries based on wild stocks of giant clams. This leads to consideration of farming (mariculture) giant clams as an alternative to fishing.

Techniques for rearing giant clams are well established and published in manuals (e.g. Heslinga *et al.* 1990; Braley 1992; Calumpong 1992). The rearing process includes phases similar to those for other cultured bivalve molluscs (Figure 5.2; Box 5.3).

Phases 1 to 3 are accomplished on land in a hatchery/nursery facility. Phases 4 to 5 involve shallow marine environments (Figure 5.3). They are the culture phases that are most crucial to sustainable exploitation and conservation of wild stocks. This is because the ocean-cultured clams may serve as a substitute for exploiting or as a means of restocking wild populations.

5.4.2   Mariculture rationale

The levels of survival of the early developmental stages of giant clams (and most other marine animals) are extremely low (Figure 5.4). The millions of

---

Box 5.3   Phases in the mariculture of giant clams

1.   Broodstock preparation and spawning induction

2.   Hatchery phase – larval rearing, settlement and metamorphosis

3.   Nursery phase – rearing juveniles in outdoor tanks (from 0.2 mm to 2 cm shell length)

4.   Ocean-nursery phase – rearing juveniles in protective containers in the ocean (from 2 cm to 10 cm shell length)

5.   Grow-out phase – culture without protection in the ocean (approximately 10 + cm shell length)

Figure 5.3 Ocean culture of giant clams in cages (ocean-nursery culture) in Solomon Islands. (Photograph from ICLARM-CAC.)

eggs soon become thousands and then hundreds of survivors within a matter of days. Many eggs are not fertilized; eggs and larvae are preyed upon by planktonic carnivores, or larvae may starve from insufficient phytoplankton; when the pediveliger larvae are ready to settle and meta-morphose, they may not be carried across a suitable reef environment; juvenile clams have to rely on being cryptic as their defence against most reef predators and most are consumed. Govan (1992) listed known predators and parasites of giant clams: at least 45 species, e.g. trigger fishes, octopods, crabs and predatory snails. It is only after several years that the juvenile clams' size and shell thickness become a significant defence against the larger predators.

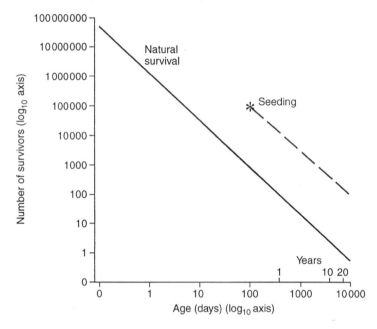

Figure 5.4 A hypothetical representation of survival versus time of the progeny of a giant clam spawning in the field. (In reality the rate of decline in numbers is likely to be more irregular.)

Mariculture of giant clams seeks to vastly improve survival through these early development stages. It does so by promoting high fertilization rates of eggs, favourable and predator-free environments for larval development, and protection for the juvenile clams until they reach sufficient size to be able to resist most predators.

Mariculture, like other farming, is usually a commercial activity involving the efficient production of primary products for markets. It is not usually about conservation or sustainable exploitation of wild resources. Most giant clam mariculture is a commercial activity. However, there are two means by which mariculture of giant clams can be used for conservation and sustainable exploitation. The first is by supplying the demand for giant clam products and therefore reducing or removing the need to exploit wild stocks. The second is by restocking wild populations with mariculture-produced juveniles.

There are a number of Pacific countries involved in giant clam farming. Some (e.g. Australia, Marshall Islands and Western Samoa) involve private companies purely with commercial motives. Others involve government or government-funded organizations (e.g. American Samoa, Cook Islands, Fiji,

FSM, Palau, the Philippines, Japan, Solomon Islands and Tonga). The latter group, while having commercial motives, may also be involved in conservation and restocking.

5.4.3 Mariculture as a substitute for fishing wild stocks

A development of giant clam farming in the Solomon Islands involves supplying village people with hatchery-produced juveniles. After a period of farming, the clams are bought back from the village farmers for sale by the organization running the hatchery. In this way the village people obtain an income from giant clams without needing to interfere with their local, wild stocks.

This innovative programme is operated by the International Centre for Living Aquatic Resources Management (ICLARM) Coastal Aquaculture Centre. The hatchery/nursery facility is based on the shore not far from Honiara (Figure 5.5). It consists of a substantial number of outdoor tanks for larvae and juvenile rearing. All six species of giant clam in the Solomon Islands have been spawned. For instance, ten batches of clam larvae and juveniles, consisting of five species, were reared during January–November 1993 (Oengpepa 1993). In the three years 1993–1995, an average of about 170 000 juvenile clams were produced for distribution to farmers from a

Figure 5.5 International Center for Living Aquatic Resources Management's Coastal Aquaculture Centre, near Honiara, Solomon Islands. (Photograph from ICLARM-CAC.)

tank area of about 240 m² (Gervis *et al.* 1995). The juveniles are reared to ocean-nursery size (about 20 mm shell length and 6+ months old) and then supplied to village farmers on a credit basis. The farmers are charged according to the age of the juveniles, reflecting the hatchery/nursery cost of producing them. Batches of juvenile clams are delivered to the farmers during monthly visits to their village.

As a result of preliminary research by ICLARM-CAC, appropriate farming sites, culture structures (protective cages, etc.) and culture protocols have been developed. Suitable farmers are selected, trained in the culture techniques, supplied with juvenile clams on credit and visited monthly with an Extension service. This service monitors their farming and gives advice with problems. A typical clam farm of a village farmer consists of a series of cages containing hundreds of clams. These cages of clams are kept in shallow water near the village (Figure 5.3). They need to be inspected several times a week to remove predators, especially predatory snails.

Each farmer is required to keep an inventory of their giant clam stocks and these are also monitored by the ICLARM-CAC for availability in relation to market demands. This means that ICLARM-CAC can draw upon appropriate stocks when they are required. Then when some of a farmer's clam stocks are sold, the farmer receives the selling price, minus the initial credit, charges for delivery and Extension services, and a small commission for ICLARM-CAC to cover operating costs.

Hambrey and Gervis (1993) gave an economic assessment of this operation, based on large *T. gigas* for meat. This was not positive and they recommended production of smaller clams and consideration of the marketability of the shell. Subsequently, the operation has shifted towards producing small clams for the aquarium trade (Bell *et al.* 1996b). For this trade, it is not unusual for a farmer to sell clams back to ICLARM-CAC after only 2–3 months of culture in the village farm. The batch of clams from one spawning will usually be sold to the aquarium trade within a year of supply to the farmers.

Recently, a net return of US$180 per typical cage has been obtained for *T. gigas* juveniles grown for the aquarium trade (Bell *et al.* 1996b). This gives a return of US$1.45 per hour, based on four cages stocked with approximately 400 juveniles and 12 hours of husbandry per week for 10 months. This clam farming has the advantage of giving the farmer time to pursue other traditional activities. By comparison, full-time labour for cocoa and copra production gives returns of US$0.38 and US$0.25, respectively, per hour (Bell *et al.* 1996b).

ICLARM-CAC (and other commercial operations) is actively seeking to develop international markets for clam products. Potential products include: adductor muscle and mantle meat, whole small clams for sashimi, marine aquarium specimens (especially the species with brightly coloured mantles, e.g. *T. crocea*), shells and shell artefacts (jewellery, dishes, etc.), hatchery products (late-stage larvae and early juveniles), and biological specimens. If the international trade in cultured giant clam products is to flourish, it must develop a broad range of products. More effective policing of the illegal international trade in wild-caught giant clam meat, shells and aquarium specimens would assist the mariculture industry. Illegal products obtained from wild resources would become more scarce, forcing prices up and making mariculture more economic.

5.4.4 Restocking reefs

A more direct use of mariculture for the conservation and sustainable exploitation of wild populations of giant clams is to restock, or 'seed', suitable reef environments with cultured juveniles from a hatchery/nursery. The rationale is that, by rearing the early development stages in culture with much higher survival, much larger numbers will persist when put out in the coral reef environment (Figure 5.4). However, it is not this simple.

Proposals for restocking various fisheries with hatchery-produced juveniles have been considered for many decades. A number of stock enhancement programmes for commercial invertebrates and fishes have been undertaken, at least on a pilot scale. The invertebrates tested include abalone (*Haliotis* species), clawed lobsters (*Homarus americanus* and *H. homarus*), Queen conch (*Strombus gigas*) and trochus (*Trochus niloticus*). In few if any cases is there clear evidence that such programmes have enhanced the fishery. For instance, the two species of clawed lobster have been cultured and released into the field in North America and Europe in various programmes since the nineteenth century, as yet without clear results (Addison and Bannister 1994). Hatchery-reared individuals may be 'naive' in terms of the skills needed to survive in the natural environment. For instance, hatchery-reared Queen conch juveniles seeded into the field were found to be inferior in morphology, physiology and behaviour compared with wild animals, leading to reduced survival (Stoner 1994).

Fortunately, 'naive' behaviour is not a problem for sedentary animals like giant clams. However, surviving juveniles in the field may be surviving because of some inherited 'fitness' factor. Probably even more important, the juvenile clams that are surviving in the reef habitat are surviving because of their locations. They have been fortunate to settle in habitats where they are very cryptic, and so far overlooked by predators, while being exposed to essential sunlight.

Individually placing large numbers of juveniles in habitats that conceal them, yet expose them to good light levels, is impractical. An alternative, scattering the juveniles in the reef environment, will result in almost total mortality: they are readily located and provide a feast for predators. There are two further alternatives. One is to provide protective structures for the juveniles as they are put out into the reef environment (as for the ocean-nursery phase of mariculture). The other is to rear the juveniles in the hatchery/nursery for several years until they reach a size where they are almost free from predation. The problem with these final two alternatives is their cost.

Leung *et al.* (1994) estimated the cost of producing year-old juvenile clams from hatcheries to be US$0.76–0.82 each in US-affiliated Pacific islands. The cost for 2-year-old clams reared initially in a hatchery/nursery and then transferred to cages in the field (ocean-nursery) increases to US$1.23–3.40 each. The cost for 5-year-old clams increases to about US$9 per clam where a floating cage system is used; however, it should not be as expensive as this where the clams are caged on the substrate. Protective cages should no longer be necessary after about 3–4 years.

No estimates are available for the cost of the alternative: rearing giant clams for a number of years in an onshore nursery until they reach a size where they can be put out in the field without suffering much predation. They will probably need to be at least 3 years old and the cost will be considerably greater per clam than where they are reared using the usual combination of hatchery/nursery and ocean-nursery system. Large amounts of tank space and turnover of water are required to sustain good growth of 1+ year-old clams. More labour is required to maintain the equipment and to manage the clams in tanks. Thus, the capital, power (pumping seawater) and labour costs will be greater per unit clam in this approach. It is not possible to give a figure for the cost of each giant clam in a reef restocking programme. There will be local factors that affect the cost, e.g. techniques used, labour cost levels, predator pressure, remoteness of restocked reefs. However, from the production cost data above, the cost is likely to be at least US$5 per seeded clam. A significant restocking programme will involve thousands of giant clams and will thus require a substantial budget.

There is a further factor to consider in restocking programme and this is maintaining the genetic diversity of the wild stocks (Munro 1993). If large numbers of hatchery-produced juveniles from a few parents are released into the field, they may overwhelm the gene pool of the wild stock with their limited genetic diversity. To reduce the chance of this happening, restocking programmes should involve a substantial number of broodstocks as the source of 'seed' clams. There are local differences in giant clam populations. Thus, local clams should be used as the source of broodstock, to avoid 'smearing' the genetic differences, unless the local stock is extinct.

5.4.5 Examples of restocking reefs

In the Solomon Islands, giant clam farmers are required by an agreement with ICLARM-CAC to set aside 20 randomly selected individuals from each batch of clams supplied to them. These are to be grown in the protective cages to a size where they can be put out on the reef with a low chance of predation (Bell *et al.* 1996a). In this way, farming of giant clams and a modest level of restocking are complementary activities and the major cost of restocking is borne by the farmers.

Silliman University Marine Laboratory on Negros Island, Philippines, undertook a reef restocking programme using juvenile giant clams in cages. More than 20 000 clams of seven species were put out at 25 sites over the period 1985 to 1992 (Calumpong and Solis-Duran 1993). About 10% survived until 1992, despite the clams being caged against predators. This was due to a combination of physical, biological and human factors, with the largest single cause of mortality (35%) being storms. As a result, Calumpong and Solis-Duran recommended careful site selection in regions that are subject to storms (cyclones/typhoons) and careful monitoring of the restocking clams. Similar problems with storm damage to giant clams in protective cages and restocked clams have been experienced on the Great Barrier Reef, Fiji and Western Samoa (Lucas 1994).

Kakuma (1989) and Murakoshi (1987) conducted reseeding experiments with *T. crocea* in southern Japan. *T. crocea* burrows into hard, reef substrates and the experiments included placing juveniles into drilled holes. Survival over 3 years varied from less than 0.5% for clams placed free on the substrate to 56% for clams placed individually in holes. Kakuma (1989) suggested that further protection could be provided by stapling mesh over the juvenile clams inserted into prepared holes, but this would be even more labour-intensive and costly.

These experiences reinforce the point made in the previous section that giant clam restocking programmes on a large scale will not be cheap and nor, it seems, will they be uniformly successful. However, with suitable reef locations and with appropriate technical input and maintenance, survival rates of more than 50% have been achieved for restocked clams (Kakuma 1989; Calumpong and Solis-Duran 1993).

5.5 SUSTAINABLE EXPLOITATION

5.5.1 Giant clam mariculture is environmentally sound

A notable feature of farming giant clams is the apparent absence of deleterious environmental effects. This contrasts with some situations where other bivalve molluscs are cultured at high densities, e.g. oyster beds and mussel rafts. There can be problems of accumulating wastes on the substrate beneath due to the 'rain' of faeces from the bivalves. The fine accumulating wastes can cause the substrate to become oxygen deficient and kill many of the normal substrate inhabitants. The problem is prevented or managed by not having excessively high densities of bivalves in culture, using areas with strong currents to carry away the faeces and shifting farming sites when the substrate starts to go 'off'.

Figure 5.6   Culture of large giant clams without protection (grow-out culture) in the intertidal zone at Orpheus Island, Great Barrier Reef.

However, there is no accumulation of faeces from giant clams, even where the clams are cultured at high densities (Figure 5.6). The faeces from giant clams are packed with algae, including waste zooxanthellae. The faeces are apparently nutritious because they are immediately consumed by the small, plankton-feeding fishes that take up residence throughout the clam farm.

Coral reefs are considered to be fragile ecosystems and farming giant clams offers an environmentally friendly culture system for exploiting this environment.

5.5.2   The need for management measures

Mariculture of giant clams has potential for being a commercial operation while serving to conserve wild stocks. Village-based giant clam farming is feasible, as demonstrated by the ICLARM-CAC programme in the Solomon Islands. It is possible to use this as a substitute for exploitation of wild stocks of giant clams on coral reefs in the region. However, this will only succeed as a substitute if the village people are persuaded not to resume harvesting the wild stocks at unsustainable levels. The same condition applies to restocking reefs with cultured giant clams. This will be a wasted exercise unless there are limitations on fishing effort in the restocked areas to keep to a sustainable level of exploitation. The alternative is that these areas will be the favoured areas for clam harvesting, because they will have the highest abundances of giant clams, at least for a while.

This problem was demonstrated in the Kingdom of Tonga where scarce adult clams were aggregated near a hatchery, to be used as broodstock, and clams were aggregated at other sites to enhance their reproduction. All the clams in several of these aggregations were illicitly harvested. Then, when a grow-out site was established near a somewhat urbanized area at Sopu, Tongatapu, there was poaching of cultured juveniles when they reached sufficient size. It was necessary to enclose the grow-out site with a barbed-wire fence and construct a watch house out on the reef-flat adjacent to the site. A guard is stationed in the watch house at appropriate times.

Ownership of Tonga's reefs and lagoons has been totally vested in the Crown since the late nineteenth century, taking away traditional and customary rights of local villages. Thus, poaching occurred where there was no local 'ownership' of the reef resources. A more favourable result has been achieved at 'Atata island, offshore from Tongatapu, where the villagers have taken responsibility for protecting their juvenile clams in an ocean-nursery near their village. They still show a sense of informal 'ownership' of reef resources. Clearly, the social environment as well as the marine environment must be considered when establishing an ocean site for culturing or restocking giant clams.

5.6 CONCLUSION

Mariculture can potentially serve as a means to conserve wild stocks of giant clams and be a means to restoring depleted wild stocks for sustainable use. This is by:

- Serving as a source of commercial giant clam products in lieu of harvesting from the wild.
- Supplying juvenile clams for restocking reef populations. This can be by focused programmes, or by dual programmes, such as in the Solomon Islands, where farming and restocking are complementary activities.

However, neither of these measures will be successful by themselves. There also needs to be effective management of the wild stocks of giant clams for sustainable use. McKoy (1980) reviewed the declining giant clam stocks of Tonga and considered a series of measures to reduce the fishing pressure and conserve the clams. These measures included closed reef areas, closed seasons, catch quotas, minimum size, prohibition of scuba and hookah diving, prohibition of power boats, export prohibitions, and licensing clam fishers (see Box 5.2). These types of measures must be implemented to complement employing mariculture as a means to sustainable use of wild stocks of giant clams.

ACKNOWLEDGEMENTS

Support to the author from the Australian Centre for International Agricultural Research for two major research projects on giant clams is thankfully acknowledged. I am grateful for assistance from ICLARM-CAC.

REFERENCES

Addison, J.T. and Bannister, R.C.A. (1994) Re-stocking and enhancement of clawed lobster stocks: a review. *Crustaceana*, **67** (2), 131–55.

Bell, J.D., Hart, A.M., Foyle, T.P., Gervis, M. and Lane, I. (1996a) Can aquaculture help restore and sustain production of giant clams? *Proceedings of Second World Fisheries Congress*.

Bell, J.D., Lane, I., Gervis, M., Soule, S. and Tafea, H. (1996b) Village-based farming of the giant clam, *Tridacna gigas* (L.), for the aquarium market: initial trials in Solomon Islands. *Aquaculture Research*.

Braley, R.D. (ed.) (1992) *The Giant Clam: a Hatchery and Nursery Culture Manual*, ACIAR Monograph No. 15. Australian Centre for International Agricultural Research, Canberra.

Calumpong, H.P. (ed.) (1992) *The Giant Clam: An Ocean Culture Manual*. ACIAR Monograph No. 16. Australian Centre for International Agricultural Research, Canberra.

Calumpong, H.P. and Solis-Duran, E. (1993) Constraints in restocking Philippine reefs with giant clams, in *Biology and Mariculture of Giant Clams* (ed. W.K. Fitt). ACIAR Proceedings No. 47. Australian Centre for International Agricultural Research, Canberra, pp. 94–8.

Carleton, C. (1984) *Miscellaneous Marine Products in the South Pacific: an Overview of Development in Member Countries of the Forum Fisheries Agency*, FFA Report 84/3. South Pacific Forum Fisheries Agency, Honiara, Solomon Islands.

Dawson, R.F. and Philipson, P.W. (1989) The market for giant clams in Japan, Taiwan, Hong Kong and Singapore, in *The Marketing of Marine Products from the South Pacific* (ed. P.W. Philipson). Institute of Pacific Studies, University of the South Pacific, Suva, pp. 90–123.

Gervis, M., Bell, J., Foyle, T., Lane, I. and Oengpepa, C. (1995) Giant clam farming in the South Pacific, past experience and future prospects. Paper presented to the Regional Aquaculture Workshop on the Present and Future Research and Development in the South Pacific Countries, Ministry of Fisheries, Tonga, November, 1995.

Govan, H. (1992) Predators and predator control, in *The Giant Clam: an Ocean Culture Manual* (ed. H.P. Calumpong), ACIAR Monograph No. 16. Australian Centre for International Agricultural Research, Canberra, pp. 41–9.

Hambrey, J. and Gervis, M. (1993) The economic potential of village-based farming of giant clams (*T. gigas*) in the Solomon Islands, in *Biology and Mariculture of Giant Clams* (ed. W.K. Fitt), ACIAR Proceedings No. 47. Australian Centre for International Agricultural Research, Canberra, pp. 138–46.

Heslinga, G.A., Watson, T.C. and Isamu, T. (1990) *Giant Clam Farming*. Pacific Fisheries Development Foundation (NMFS/NOAA), Honolulu.

Juinio, A.R., Menez, L.A. and Villanoy, C. (1987) Use of giant clam resources in the Philippines. *Naga*, **10**, 7–8.

Kakuma, S. (1989) *Tridacna crocea Re-seeding Handbook*. Okinawa Fisheries Extension Office, Okinawa. (In Japanese.)

Leung, PS., Shang, Y.C., Wanitprapha, K. and Tian, X. (1994) Production economics of giant clam (*Tridacna*) culture systems in the U.S.-affiliated Pacific Islands, in *Economics of Commercial Giant Clam Mariculture* (eds C. Tisdell, Y.C. Shang and PS. Leung), ACIAR Monograph No. 25. Australian Centre for International Agricultural Research, Canberra, pp. 267–91.

Lucas, J.S. (1994) The biology, exploitation and mariculture of giant clams (Tridacnidae). *Reviews in Fisheries Science*, **2** (3), 181–223.

McKoy, J.L. (1980) *Biology, Exploitation, and Management of Giant Clams (Tridacnidae) in the Kingdom of Tonga*. Fisheries Bulletin No. 1, Fisheries Division, Nuku'alofa, Tonga.

Munro, J.L. (1993) Strategies for re-establishment of wild giant clam stocks, in *Genetic Aspects of Conservation and Cultivation of Giant Clams* (ed. P. Munro). ICLARM Conference Proceedings, Manila, pp. 17–21.

Murakoshi, M. (1987) Farming of the boring clam, *Tridacna crocea* Lamarck. *Galaxea*, **5**, 239–54.

Oengpepa, C. (1993) Recent spawnings at the CAC. *Clamlines*, **12**, 9.

Pearson, R.G. (1977) Impact of foreign vessels poaching giant clams. *Australian Fisheries*, **36** (7), 8–11, 23.

Stoner, A.W. (1994) Significance of habitat and stock pre-testing for enhancement of natural fisheries: experimental analyses with the Queen conch *Strombus gigas*. *Journal of the World Aquaculture Society*, **25** (1), 155–65.

Wells, S.M., Pyle, R.M. and Collins, N.M. (1983) Giant clams, in *The IUCN Invertebrate Red Data Book*. IUCN, Gland, Switzerland, pp. 97–107.

Yonge, C.M. (1975) Giant clams. *Scientific American*, **232**, 96–105.

6

Exploitation and conservation of butterflies in the Indo-Australian region

Tim R. New

6.1 INTRODUCTION

The amount of protein in a single butterfly is small, yet the discarded bodies of butterflies whose wings are used in Taiwan to construct tourist items such as laminated tablecloths and placemats are sufficiently numerous to be used for pig food. But, unlike most of the other taxa discussed in this book, the major commercial appeal for butterflies is not as human food or manufacturing commodities but, simply, aesthetic. Butterfly collecting is a popular hobby and, as with rare stamps and other 'collectables', rare butterflies can command very high prices – either legally or on a black market. In the past, a number of professional collectors in the tropics made good livings from selling specimens to wealthy collectors or patrons, and the butterflies of the western Pacific and southeast Asia have long held special fascination for collectors. Lord Walter Rothschild, who built up the largest-known private collection of Lepidoptera, containing some 2¼ million specimens, employed more than 400 collectors of butterflies and/or birds in various parts of the world. One of the major areas of concentration for collectors has been the Indo-Pacific (Figure 6.1 – after Rothschild 1983), reflecting that many of the rarest, largest and most spectacular butterflies occur there, many of them very locally, and the faunas are highly diverse. Some of the early collectors tell of shooting large, high-flying birdwing butterflies with dust-shot, or of employing local people to stun them with blunt arrows, and the lengths which were pursued to obtain such highly valued species make fascinating reading (e.g. Meek 1913).

However, collectors put a premium on first-class specimens and, even though they may be very rare, butterflies with worn or tattered (or, even,

Conservation and the Use of Wildlife Resources. Edited by M. Bolton.
Published in 1997 by Chapman & Hall. ISBN 0 412 71350 0.

Figure 6.1 Places where Lord Rothschild's collectors operated in the Indo-Pacific region, mainly seeking birds and butterflies (after Rothschild 1983).

shot-holed!) wings generally are not as desirable as fresh, undamaged specimens. 'Quality control' is an important parameter of butterfly trade, and recent moves towards captive rearing (farming) or enhancement of habitat to increase breeding densities of desirable species for harvest (ranching) are designed, in part, to furnish first-class specimens for trade at a lower cost and greater sale value than captured individuals (Orsak 1993). World trade in butterflies, much of it based on the fauna of the Indo-Australasian region, may be as high as US$100 million/year (Parsons 1992a), but exact figures on the volume and value of the butterfly trade are extremely difficult to obtain.

The conflict between overcollecting as a putative threat, and conservation, is perhaps nowhere more emotional than for rare tropical butterflies in the western Pacific and south-east Asia. Much protective legislation for other insects and other groups of invertebrates has developed directly from butterfly examples, where estimates of status lead to (sometimes, overzealous) legal prohibitions on collection and export of potentially lucrative taxa. Such prohibition is only rarely accompanied by more constructive measures to conserve the species and the habitats on which they and myriad less conspicuous organisms depend. Collecting in itself is only rarely a threat to butterfly species, and the major threatening process is undoubtedly the destruction of habitat.

6.2 EXPLOITATION AND CONSERVATION

Realization that certain desirable butterflies can be harvested in ways that:

- provide substantial income to the practitioners
- reduce collector pressure on vulnerable or small wild populations, and
- help to reduce other human intrusions into primary forests and other habitats

is now leading to management utilizing indigenous butterflies as potent umbrella and flagship taxa for wildlife conservation and sustainability of the species and their habitats in parts of the western Pacific and nearby areas. The major impetus for development of butterfly ranching has come from work in Papua New Guinea (reviewed comprehensively by Parsons 1992a) which has evoked global interest and spawned emulative operations in several other countries. Much of this activity was not focused primarily on conservation in the early stages of its development but some more recent projects depend more comprehensively on sustainability of natural habitats and species assemblages. 'Insect farming is at last gaining recognition as a viable environmentally sound enterprise, particularly appropriate for adoption in developing equatorial countries around the world' (Clark 1992). Butterflies as a self-renewing, sustainable resource can be integrated into promoting people's livelihoods, by providing financial reward linked with important conservation components, and can play an important role in the conservation/development dilemma so prevalent in many parts of the region.

6.3   BUTTERFLIES AND MARKETS

Much of the focus of ranching operations is, naturally, on taxa which are rare and are otherwise difficult to obtain in numbers. However, much of the butterfly trade is in common, showy species – what Collins and Morris (1985) termed the 'high-value/low value' trade – and in livestock for exhibition in the burgeoning number of butterfly houses throughout the world. Most such butterflies are captured in the wild, but constitute a vital part of the overall butterfly resource on which commercial operations depend. Common species command only low prices, but their bulk can result in important financial return for field collectors, and the markets for them (predominantly for artwork and tourist items, as well as for collectors) seem assured. Clark (1992) also noted that, unless a centralized agency will handle the low-return common species, local collectors may get discouraged, and the agency will thereby miss out on the 'odd rare item' for which the profit margin will be much higher.

Much of the emphasis in butterfly trade is on members of the family Papilionidae (the swallowtails, birdwings and their allies), many of which are large, spectacular and rare. Most of the approximately 570 species are listed in dealers' catalogues, and the rarest birdwings (*Ornithoptera* spp.) are the largest and most highly prized of all butterflies. These are restricted to the western Pacific region, and the greatest regional diversity of Papilionidae is also in this area; Indonesia and the Philippines together support nearly 150 members of the family (Collins and Morris 1985). The importance of several birdwing species was acknowledged by the Papua New Guinea government in 1967 by their designation of seven forms (six species and the hybrid form known as *O. allotei*) as protected and 'national butterflies'. Several of these species are known now to be not as severely threatened as supposed earlier (Parsons 1992a), but the Papua New Guinea legislation (with acknowledgement of butterfly protection in the national constitution) was a major impetus to development of butterfly ranching, of controlling collecting and trade of butterflies, and as the forerunner of practical conservation measures centred on habitat enrichment and sustainability.

6.4   INTERNATIONAL TRADE ISSUES

The designation was important also in leading to the listing of all birdwing butterflies on the Convention on International Trade in Endangered Species (CITES), a step which was taken in part for convenience (as many species are very similar in appearance and some very common species are difficult for a non-specialist to differentiate from highly threatened rare taxa), but which imposes the need for permits for export and import of specimens and thereby allows for monitoring of the amount of trade. This need for permits therefore applies even to abundant taxa (such as *Ornithoptera priamus* and *Troides oblongomaculatus*) which are thereby subject to international regulation whilst not receiving (and, indeed, not needing) any national or more local protection. Most birdwings are on CITES Appendix II (in which trade is monitored in order to avoid threats of extinction) whereas *Ornithoptera alexandrae* is on Appendix I (in which wild-caught specimens cannot be traded except under very unusual circumstances). The 'blanket protection' of birdwings received mixed reactions (Parsons 1992a). Greater control of trade can come through centralized operations.

6.4.1 Trade strategy in Papua New Guinea

In Papua New Guinea, the Insect Farming and Trading Agency (IFTA), now part of the commercial branch of the PNG University of Technology, Lae, was founded to promote the rational use of the country's insect resources, with a trade based primarily on butterflies (National Research Council 1983; Mercer and Clark 1989; Clark and Landford 1991). The agency was situated at Bulolo, providing for ready access to a wide range of butterfly habitats over mid to high latitudes, and has the following roles:

- Quality control – provision of high quality specimens with data.
- The sole exporter of insects in Papua New Guinea. Stock is purchased from a network of collectors and ranches (some 800 people, spread widely through the country, supply the agency with specimens in return for financial reward).
- Research and monitoring of butterfly species and promoting needs for conservation.

These are functionally linked and form the core of an important strategy for sustainable use of wildlife and habitats.

The combination of need for first-class specimens of butterflies and other insects, with the continued demand for these by collectors in many countries, and for control of marketing, provided ample opportunity to involve local people in many parts of the country. The major reasons for establishing IFTA (summarized from Parsons 1992a, and enlarging on the above) were:

- To promote the production and sale of butterflies as an alternative source of income for subsistence farmers, especially in less advanced areas of the country.
- To restrict trade in insects to PNG citizens.
- To ensure that fixed/reasonable prices are paid to collectors and farmers and to assure expedient payment for butterflies and other insects.

- To provide a centralized body as a communication centre for sellers and purchasers, and to serve as an official agent for business overseas buyers.
- To ensure the highest possible quality of stock, including locality data for specimens.
- To act as an educational centre for instruction in insect farming and trading methods, and to provide basic equipment for participants.
- To ensure that insects are treated as a renewable resource.
- To promote the conservation of butterflies and their habitats.

6.5 BUTTERFLY RANCHING

The basis of butterfly ranching is local habitat enrichment, undertaken on a formal garden plot or a more extensive basis. A farming manual from IFTA (Parsons 1992) instructed PNG citizens in establishing butterfly gardens, emphasizing the importance of rearing high quality specimens and protecting the stock from predators and parasites. The farm may be only around 0.2 ha in area and can be surrounded with hedges of nectar-bearing plants as an adult food resource to encourage butterflies, and exclude animals such as pigs. Up to 500 plants of the common vine *Aristolochia tagala* can be planted in this area to grow up the branches of shade trees, such as *Leucaena*, which can also be selected as foodplants of other insects of value to IFTA and which can therefore be collected and sold. The vines are the foodplant of caterpillars of several birdwing species, and female butterflies attracted to their vicinity by the nectar plants oviposit on them.

6.5.1 Methods of harvesting large caterpillars

Pupae can be collected daily from an established farm. The pupae darken near hatching, and the adult wing and body colours become visible through the case. The harvested pupae, attached to a leaf or stem, are transferred to an emergence cage, in which they are pinned head uppermost to a net or board. Care may be needed to protect pupae from attack by rats or ants: for example, the emergence cage may be supported on legs placed in bowls of water to hamper ants climbing up. Pupae are kept in a shady place, and sprayed lightly with water two or three times a week to prevent desiccation (National Research Council 1983).

Large caterpillars and pupae are collected and caged in due course, and up to about half are left to constitute field populations which, hopefully, will serve to sustain the crop. Newly emerged birdwing butterflies are allowed to expand their wings fully and 'harden', after which perfect specimens for marketing are held carefully by the thorax and killed by injection of boiling water or ethyl acetate. The dead butterflies, with wings closed, are placed carefully in triangular greaseproof paper envelopes and dried in the open air for several days – again being protected from predators and adverse weather. Once dried, the envelopes containing the specimens are stored in airtight boxes and when sufficient individuals have been collected they are packed in cardboard boxes, with a small amount of naphthalene to prevent insect attack, for transmission to IFTA. Data for each specimen can be written on the envelope, as many specimens are sold to overseas collectors in this 'prepared' state.

Formal farms are rather rare, and many villagers simply plant the vines more extensively in and around the village, wherever there are suitable support trees. The general appearance of many butterfly farms is therefore that of secondary vegetation rather than any defined area. Nevertheless, the planting of *Aristolochia* vines in secondary forest fosters increase in birdwings, possibly without an equivalent increase of natural enemies which may result from the more crowded conditions of formal garden ranching.

Income from a butterfly ranch in Papua New Guinea can easily approximate that of a minimum rural wage, to which only a low proportion of citizens have access because of high unemployment, and can be several-fold higher than from insect collecting alone. However, a survey of incomes by Sekhran and Miller (1995) showed that only 86 of 836 farmers/collectors had an annual income of 1000 kina or more, whereas 400 earned less than 200 kina/year. Much higher rewards are documented: for example, up to US$14 000 for a rancher of two rare island butterflies (Orsak 1993). But the ability to harvest butterflies for sale on a regular basis has important ramifications for people's way of life and for conservation. First, they demonstrate a tangible reward from a forest product whose sustainability will depend on continued availability and management of the habitat. Second, by providing the means to purchase food and other commodities, they reduce the need to clear native vegetation for agriculture and provide additional means to protect natural habitats. Third, the population of naturally sparsely distributed or scarce butterflies may be enhanced through husbandry, so that the return is demonstrably greater than from simple collecting, and there is some 'built-in' protection for rare species by harvesting only a proportion of the population, and by not capturing damaged individuals which are thereby left as breeding stock.

These principles led New and Collins (1991) to suggest that controlled exploitation based on ranching merits serious appraisal as a powerful conservation tool and incentive for many rare or commercially desirable species of swallowtail butterflies (see also New 1994; New *et al.* 1995).

6.6 QUEEN ALEXANDRA'S BIRDWING: A FLAGSHIP SPECIES FOR CONSERVATION

The world's largest butterfly has been reduced, by destruction of its habitat associated with oil-palm plantation and timber extraction, to small isolated populations in remnant forest patches in the region of Popondetta, PNG. It is a species of primary forest and tall secondary lowland forest, and destruction of this specialized vegetation type for industrial purposes has been compounded by greater numbers of people moving into the area because of the likelihood of continuing employment. Other areas are not under as great a threat. The history of conservation neglect of this remarkable insect is documented by Parsons (1992b), this despite widespread demonstration more than a decade ago of the need for urgent conservation measures to be taken. Based, in part, on a detailed management plan by Orsak (1992), the Australian and Papua New Guinea governments are currently pursuing conservation of the butterfly as part of a wide-reaching scheme of habitat conservation and rational development in the Northern (Oro) Province.

One important aspect of this programme is to create economic incentives for conservation, including facilitating ranching of butterflies. It is hoped that trading in *O. alexandrae* may evolve as populations increase, drawing on the basic strategy of 'strengthening any traditional conservation tendencies' of landowners (AIDAB 1993). To this end, it is proposed to implement butterfly ranching at selected sites which contain significant habitats conserved for *O. alexandrae*. Activities are planned to include:

- training in ranching techniques and education to enhance the understanding by practitioners
- supervision of ranching and sustainable collecting of other insect specimens, integrating IFTA's quality control practices with the activities of local people and with a substantial training component for local people and local staff
- marketing the specimens effectively

These activities are viewed as one major output of the project component focused on conservation of primary areas; the other activities projected are the selection of prime conservation areas and establishment of these, so that ranching activities can be undertaken in areas which are already recognized as biologically significant and which are protected adequately from unplanned intrusions. It is not yet clear whether *O. alexandrae* is amenable to ranching and there are legal problems with promoting it in trade. For example, only individuals of CITES I listed species which have been bred in captivity can be traded legally under permit, and they are then equated to CITES Appendix II taxa. 'Ranching' is not recognized as equivalent to 'captive breeding' by CITES and some countries (including Australia), so that regional trade is restricted.

The broad assumptions of the international project include:

- that conservation can be adopted into local concepts of development sufficiently strongly to secure enough primary habitat for the butterfly to survive

- that surveys and research can demonstrate *O. alexandrae* population needs with sufficient confidence to guide resource management to enhance the butterfly's survival
- that controlled trading of the butterfly will be instituted in a manageable fashion through IFTA

Queen Alexandra's birdwing, reflecting its importance as local faunal emblem pictured on the Oro Province flag, has thus become a significant flagship species for conservation (New 1991a). Protection of the butterfly should be achieved by conservation of habitat, including primary areas and sales of a controlled supply of the butterfly derived from ranching, together with sales of other species reflecting the demands already generated through IFTA (AIDAB 1993). Destruction of primary forest, hopefully, will be slowed and enrichment of secondary forests continue over a considerable period.

6.7 THE LIVESTOCK MARKET AND CONSERVATION

The link between 'butterfly houses' and practical conservation is commonly very tenuous (Collins 1987; New 1991b), but three aspects of their operations are directly relevant to sustainable use of butterflies:

- supplying the market in livestock from ranched or farmed stock
- educating the viewing public on the needs and practice of butterfly conservation
- knowledge gained in rearing operations on site may be applicable to related species of greater conservation concern

The major exhibits are of long-lived, large, showy species, most of them common and of little individual commercial value. In the Australian region, butterfly houses are still sufficiently novel to attract numerous visitors: that at the Melbourne Zoo, for example, is consistently among the most popular exhibits of the institution, and ones in Queensland are regular features on tourist agendas. They exhibit only native Australian species. In south-east Asia, the Penang Butterfly House (Western Malaysia) is the largest operation in the region and has substantial export trade in livestock (Goh 1987) as well as producing an informative and well illustrated guidebook (Khoo and Chng 1987). For many other butterfly houses, such educational adjuncts are woefully inadequate, although many help to enhance dead stock trade by offering tourist items such as butterfly artwork, or mounted specimens. This is itself important if the specimens have been marketed initially through agencies such as IFTA, simply because part of the cost then is returned to villagers and may help, as above, to protect butterfly habitats: part of Orsak's (1993) message that 'killing butterflies to save butterflies' is an important tool for forest conservation in Papua New Guinea.

6.8   OTHER EFFORTS TO PROMOTE BUTTERFLY TRADE

Several other projects based to varying extents on the IFTA have been initiated elsewhere in the region, with similar overall aims of promoting trade in butterflies and augmenting the livelihoods of local people. Each has to adapt to local social and political regimes, and to capitalize on local butterfly fauna, but none is yet as advanced or important as IFTA, in part because most are still in the relatively early stages of establishment or operation. Not all countries in the region may be suitable for butterfly farming. Standards of living in Taiwan, for example, have risen sufficiently that the overall butterfly trade has declined simply because people no longer need the income gained formerly from selling butterflies. However, there appears to be ample opportunity in such countries as the Solomon Islands and Indonesia, with fauna and conservation needs similar to those of Papua New Guinea. Both countries have substantial trade in dead butterflies, including local or endemic birdwings, and much of this is poorly controlled.

For the Solomons, Macfarlane (1985) commented that the birdwings could usefully be exploited for conservation, but no central agency seems to have become established. Leary (1991) noted that trade in *O. victoriae* may be controlled by government quota.

Wildlife has been traded commercially since 1987 in the Solomons, and the policy guidelines for this trade are administered by the Environment and Conservation Division of the Ministry of Natural Resources. Most of the cash return for butterflies (predominantly *O. victoriae* and *O. priamus urvilleanus*) goes to Honiara-based exporters rather than to village people, despite 'justification' for the wildlife trade as generating cash income for rural people. Ranching as yet plays little part in butterfly trade, although this may be increasing to augment gains from wild-caught material.

In Irian Jaya, Indonesia, serious attempts have been made to set up an agency in the Arfak Mountains, with major farming focus on *O. rothschildi* (endemic to the Arfak region) and *O. tithonus*; neither species occurs in Papua New Guinea, and they thus represent 'gaps' in the IFTA taxon spectrum which many collectors would like to fill. They are especially important among the species for which Indonesia would have global monopoly, and would be augmented by a number of endemic Indonesian swallowtail species (Parsons 1992a; Neville 1993). Village-level operations were viewed as a viable approach in Irian Jaya (Morris 1986), but local political conditions have hampered progress on establishing an agency through the Worldwide Fund for Nature. By 1993 more than 800 people/families had registered as either having or intending to plant butterfly farm areas, typically about 20 m square and having about 50 *Aristolochia* vines. *O. tithonus* is the predominant species harvested, and appears to be very tolerant of such 'artificial' enriched habitats (Parsons 1993).

The major problem confronting the project has been delay in granting export permits for butterflies from the Indonesian National Government, resulting in farmers losing faith in the WWF project and opting to sell their stock through local dealers, where possible, as the project is not able to purchase them. In short, the 'middle man' dealers make large profits at the expense of the individual ranchers. If these problems can be overcome, it should be feasible to eventually promote butterfly conservation effectively, as part of the broader Arfak Mountains Conservation Area Management Plan.

The last important example noted here is the Xishuangbanna area of south-western Yunnan Province, China, a tropical region bounded by Laos and Myanmar, rich in butterflies but economically among the poorest parts of China. Following a report by Mackinnon (1987) that a butterfly farming project would be feasible, Parsons (1992a) initiated a demonstration butter-fly farm (including a large flight dome and a nursery of foodplants) in the large Mengyang Forest Reserve. A central marketing agency (also modelled on IFTA), the Division of Insect Farming and Trading (DIFT), is operated through the Yunnan Forestry Bureau, and it is hoped that this will even-tually foster and coordinate butterfly farming and collecting throughout China, so that the now localized activities can be extended to promote the conservation of many desirable butterflies – mainly species of *Parnassius* (the Apollo butterflies) and *Bhutanitis* (the 'glories'). At Xishuangbanna, early successes in ranching the birdwings *Troides helena* and *T. aeacus* were followed by broadening the variety of foodplants grown to increase the number of ranched taxa, and consideration is being given also to farming some of the more spectacular silkmoths, such as *Argema maenas*. None of the butterflies listed on the schedules of Protected Insects in China is known to occur around Xishuangbanna (Parsons 1992a), but difficulties in obtaining export permits to facilitate development of DIFT's trading operations have hampered initial progress.

6.9 DISCUSSION

Trade in dead butterflies is widely regarded as anathema by lepidopterist supporters of butterfly conservation, and there is no doubt that many of the most 'desirable' (rarest, most expensive, listed on protection/collection pro-hibition schedules) species are indeed traded illegally, at inflated prices, and with little (if any) return to the people of their area of origin. Yet, dead butterflies, if emanating from controlled, centralized, collecting or ranching operations, can be an important component of practical conservation, as stressed by Orsak (1993), and there is abundant need to demonstrate and extend the positive initiatives for conservation resulting from the IFTA operation and its recent extensions in Papua New Guinea. The specimens, especially of Papilionidae, are then likely to constitute part of a controlled sustainable harvest, produced and marketed under conditions which guarantee income to the producers and promote effective conservation of the butterflies and their habitats. The prospects for commercially viable live displays (butterfly houses) over much of the region are rather limited, although eco-tourism is likely to increase over the western Pacific, and broader education of purchasers – linking the interdependence of trade and practical conservation – merits considerable attention, together with overcoming the current impediments to promoting trade.

IFTA has demonstrated amply the need for centralized control of butterfly ranching operations, to ensure the integrity of subsequent trade and to give people a powerful incentive to safeguard natural habitats on which their livelihood can be seen to depend. Linked with this, field collections of other species, which cannot be ranched easily, can also be sustained as the habitats are at least reasonably secure. However, an important difference between butterflies and most of the other wildlife discussed in this book is that data on population sizes, mortality schedules, and even basic biology are commonly fragmentary or non-existent. Some information on birdwing life cycles exists, but quantitative data are rare, and for many other species even the larval foodplants are unknown. For these, any form of interventive habitat management is premature and even levels of permissible collecting may be difficult to assess. Until sounder biological information is available, the wisdom of not taking damaged individuals (wing blemishes are unlikely to affect an individual butterfly's breeding capability) and of not over-harvesting from ranched 'butterfly crops' is obvious. These precautions are an integral part of assuring the future of butterfly ranching, anywhere in the world.

REFERENCES

AIDAB (Australian International Development Assistance Bureau) (1993). Queen Alexandra's Birdwing Butterfly Conservation Project. Feasibility Study. Canberra (unpublished document).

Clark, P.B. (1992) Organisation and economics of insect farming. Paper read at invertebrates (microlivestock) farming seminar. La Union, Philippines, November 1992.

Clark, P.B. and Landford, A. (1991) Farming insects in Papua New Guinea. *International Zoo Yearbook*, 30, 127–31.

Collins, N.M. (1987) *Butterfly Houses in Britain. The Conservation Implications.* IUCN, Gland.

Collins, N.M. and Morris, M.G. (1985) *Threatened Swallowtail Butterflies of the World.* IUCN, Gland and Cambridge.

Goh, D. (1987) Quoted in Bökemeier, R. and Soutif, M. Pillards de papillons. *Geo Magazine*, Paris, 142–158.

Khoo Su Nin and Chng, W.W. (1987) *Penang Butterfly Farm Guide Book.* Yeoh Teow Giap, Penang.

Leary, T. (1991) A review of the terrestrial wildlife trade originating from Solomon Islands. *Australian Zoologist*, 27, 20–7.

Macfarlane, R. (1985) Insect farming and trading – Solomon Islands. *Papilio International*, **2**, 127–9.

Mackinnon, J. (1987) Ecological guidelines for the development of Xishuangbanna Prefecture, Yunnan Province, China (unpublished report to WWF, Gland).

Meek, A.S. (1913) *A Naturalist in Cannibal Land*. Fisher Unwin, London.

Mercer, C.W.L. and Clark, P.B. (1989) Organisation and economics of insect farming in Papua New Guinea. Proceedings of the 1st Symposium of Papua New Guinea Society of Animal Production (Smallholder Animal Production in Papua New Guinea), pp. 62–70.

Morris, M.G. (1986) A Butterfly Farming and Trading Agency in Irian Jaya. Report to WWF and USAID on Consultancy Project 920 in Indonesia (mimeographed).

National Research Council (1983) *Butterfly Farming in Papua New Guinea*. Managing Tropical Animal Resources Series. National Academy Press, Washington, DC.

Neville, D. (1993) Butterfly farming as a conservation tool; lessons learnt during implementation of butterfly farming in the Arfak Mountains, Irian Jaya. Paper read at International Butterfly Conference, Ujung Pandang, Indonesia, August 1993.

New, T.R. (1991a) Swallowtail butterflies as flagships for insect conservation. *News Bulletin, Entomological Society of Queensland*, **19**, 95–107.

New, T.R. (1991b) *Butterfly Conservation*. Oxford University Press, Melbourne.

New, T.R. (1994) Butterfly ranching: sustainable use of insects and sustainable benefit to habitats. *Oryx*, **28**, 169–72.

New, T.R. and Collins, N.M. (1991) *Swallowtail Butterflies. An Action Plan for their Conservation*. IUCN, Gland and Cambridge.

New, T.R., Pyle, R.M., Thomas, J.A., Thomas, C.D. and Hammond, P.C. (1995) Butterfly conservation management. *Annual Review of Entomology*, **40**, 57–83.

Orsak, L.J. (1992) Saving the world's largest butterfly, Queen Alexandra's birdwing (*Ornithoptera alexandrae*). Five year management Action Plan. Scientific Methods Inc., Durham, California.

Orsak, L.J. (1993) Killing butterflies to save butterflies: a tool for tropical forest conservation in Papua New Guinea. *News of the Lepidopterists' Society* (May/June 1993), 71–80.

Parsons, M.J. (1982) *Insect Farming and Trading Agency Farming Manual*. IFTA, Bulolo.

Parsons, M.J. (1992a) The butterfly farming and trading industry in the Indo-Australian region and its role in tropical forest conservation. *Tropical Lepidoptera*. Supplement 1, 1–31.

Parsons, M.J. (1992b) The world's largest butterfly endangered: the ecology, status and conservation of *Ornithoptera alexandrae* (Lepidoptera: Papilionidae). *Tropical Lepidoptera* Supplement 1, 33–60.

Parsons, M.J. (1993) Progress review of insect farming and trading in Irian Jaya. Unpublished report to WWF Project No. 100085.

Rothschild, M. (1983) *Dear Lord Rothschild*. Balaban, Glenside, Pennsylvania.

Sekhran, N. and Miller, S. (eds) (1995) *Papua New Guinea Country Study on Biological Diversity*. Department of Environment and Conservation, Waigani, Papua New Guinea.

7

Managing the Crocodilia: an integrated approach

7.1 INTRODUCTION

Crocodilians are the only living members of the Archosauria, an ancient group of reptiles which included the dinosaurs and the progenitors of birds. Existing species, of which there are about 22 (depending on determination of subspecies) occur throughout the tropics and subtropics, and most authorities classify them, on the basis of snout proportions and differences in teeth and skull, into three subfamilies: Crocodilinae (which includes the true crocodiles); Alligatorinae (alligators and caimans); and Gavialinae, a single genus represented only by the gharial (*Gavialis gangeticus*) of the Indian subcontinent.

Significant commercial trade began in the nineteenth century when leather from American alligator (*Alligator mississippiensis*) skins was favoured for boots and saddlebags. But it was not until after the Second World War that global commerce in crocodilian skins reached its peak. During the late 1950s and early 1960s an estimated 5 to 10 million skins a year were being traded (Inskipp and Wells 1979).

The most profitable leather is made from the belly skins of those species which do not produce bony plates (osteoderms) in the skin of the ventral surface. These belly skins can be given a highly lustrous 'classic' finish (Fuchs 1975). The most valuable of all skins in trade, because of its small scale pattern as well as the absence of osteoderms, is that of the saltwater or estuarine crocodile (*Crocodylus porosus*) from the Indo-Pacific. At the other end of the market, the caimans of South America have ossifications in dorsal and ventral regions and only the skin of the flanks and throat can be given a classic finish.

Crocodile and alligator populations crashed under the pressure of hunting largely because the process of hunting was removing the least expendable

Conservation and the Use of Wildlife Resources. Edited by M. Bolton.
Published in 1997 by Chapman & Hall. ISBN 0 412 71350 0.

segment of the population. The replacement of an adult crocodilian in the wild takes several years and is likely to involve the production and loss of large numbers of eggs and hatchlings. In the 1960s and 1970s the alligator and crocodile species which had been the mainstay of the specialist tanneries were in such short supply that caiman skins had to be used to keep the tanneries operating. By the end of the 1970s not more than 1.5 million crocodilian skins in total were thought to be entering world commerce each year and caiman skins (*Caiman crocodilus*) were comprising at least two-thirds of the total trade volume. Skins of the American alligator and four crocodile species (*C. niloticus, C. porosus, C. novaeguineae* and *C. cataphractus*) made up the bulk of the remainder (Hemley and Caldwell 1984).

7.2 CROCODILIAN BIOLOGY

Some of the characteristics of Crocodilia, with importance for management and use, are relevant to this chapter.

7.2.1 Thermoregulation

Like all reptiles the crocodilians are ectotherms and are mainly dependent upon heat-seeking or heat-avoiding behaviour to maintain their body temperature. The more temperate species can tolerate low winter temperatures and the American alligator can even survive being frozen over provided it can maintain a breathing hole in the ice (Brisbin *et al*. 1982). In general, however, crocodilians can only remain healthy and active within the temperature range of 25–35°C and most species cannot tolerate a body temperature below about 15–20°C for prolonged periods. Hatchlings, at least in species which have been studied, prefer to be rather warmer than do juveniles and adults (Lang 1987) but a body temperature above 40°C was found to be lethal in American alligators (Wilbur 1960).

7.2.2 Food and feeding

Crocodilians are all predators. In general, hatchlings feed on small invertebrates and as they grow they catch a wider variety of prey. Fish may predominate in the diet of juveniles, but adults (except for a few specialist fish-eaters) often kill more mammals.

Fat is stored in the tail and elsewhere and this, together with a slowed metabolism in cooler periods, enables crocodilians to live without food for months at a time.

7.2.3 Reproduction

Copulation takes place in water and all species lay eggs. One clutch a year is the rule and clutch size varies from a few eggs to more than 80 depending on species and age of the female. Because crocodilians lack sex chromosomes, the sex of the hatchlings is determined environmentally. In all species studied, a very small difference in incubation temperature (0.5–1°C) during critical early stages of incubation determines the sex of the hatchlings (Ferguson and Joanen 1982; Ferguson 1985; Lang and Andrews 1994).

Alligators, caimans and some crocodile species build a mound nest by scraping together vegetation and soil in a heap up to about a metre high. Others excavate a hole in sand or earth, as do turtles. Eggs, in either type of repository, commonly take about three months to hatch and during this time the mother may have to fend off a variety of egg-eaters such as pigs, varanid lizards, baboons – and people.

At hatching time the mother excavates the nest and may assist in the hatching process by rolling the eggs between tongue and palate before carrying the hatchlings to the water. For the first few weeks the youngsters remain close to the mother, commonly basking on her head and back as she floats. Despite this maternal care a very large proportion of eggs are lost in the wild. Flooding of nests, as well as predation, is known to be a major factor in many localities.

After hatching, young crocodilians are eaten by a different array of predators, including storks, herons, predatory fish and larger crocodilians. Only a tiny percentage of hatchlings are likely to reach maturity.

7.3 CAPTIVE REARING

The first commercial crocodile farm was probably that at Samut Prakan, near Bangkok, established in 1950, but alligator farms existed in the USA during the 1960s and by that time there was widespread interest in trying to increase stocks by collecting eggs from the wild so that incubation and rearing could be achieved without the losses which occur in nature – a procedure that was later to become known as ranching.

In Louisiana, USA, an extensive research programme was initiated in 1964 into the biology and captive propagation of the American alligator. As the programme began with the capture and accommodation of adult alligators, it was some years before findings on egg incubation and captive rearing were available. Meanwhile, in southern Africa and elsewhere, policies and techniques for commercial rearing of crocodilians were being pioneered largely by a gradually increasing number of entrepreneurs using whatever resources and official assistance were available to them. Circumstances differed considerably and there were many failures.

7.3.1 The basic requirements

The necessities – correct temperatures, humidity, diet and not too much stress – are really quite few. A variety of management techniques have proved successful but those put forward by Joanen and McNease (1975, 1976, 1979, 1987) for American alligators can be used as a standard against which results, otherwise obtained, can be compared.

(a) Incubation

In nature, crocodile eggs are not moved during the time between laying and hatching. Webb *et al.* (1987) have offered a functional overview of crocodilian eggs which suggests that rotating an egg will be lethal if it occurs after the embryo has become attached to the shell membrane (1 day after laying) and before the allantois is sufficiently developed at about 13 days. During collection and transportation, rotation can be avoided but jolting and shaking over rough roads are not easy to eliminate.

Within the Rockefeller Wildlife Refuge in Louisiana, eggs are collected usually within 24 hours of laying. They are arranged in single layers, in natural nesting material, on trays of wire mesh. Trays are then set 8 cm above water in controlled environment chambers. Relative humidity is maintained at 90–92% and the nest material is occasionally sprinkled with water. With incubation temperatures of 31–31.7°C hatch rates of 92% are achieved.

(b) Care of hatchlings and juveniles

The floors of the controlled environment chambers were stepped so that half the width held water to a depth of 15 cm and the other half was dry concrete. The alligators remained in the chambers for 3 years. For the first 10 days after hatching the temperature was held at around 32–33°C but the grow-out temperature could be as low as 28°C. Stock density was limited to one alligator per 0.1 m² until the animals were 1 year old. At that time the density was reduced to one animal per 0.3 m². Sorting the animals according to size was thought to be beneficial by reducing bullying on the part of the fastest growing individuals – which could result in the smallest ones being deprived of food.

Routine maintenance consisted of providing food on 5 days each week for the first year and 3 days a week thereafter. Food was minced marine fish (mostly *Micropogon undulatus*) or nutria (*Myocastor coypus*) with a multi-vitamin supplement. Tanks were cleaned, and water changed, within about 15 hours of every feed.

Within these chambers alligator survival, from hatching to the age of 3 years, averaged 95%. At 3 years old the alligators were transferred to outdoor pens. At this age the average animal was 160 cm long, weighed 21 kg and had consumed about 105 kg of food (Figure 7.1). Alligators of a comparable size in the wild would be about 6–7 years old (Chabreck and Joanen 1979). For a given length, wild alligators were about 10% lighter than those reared in heated enclosures (Coulson *et al.* 1973).

Faster growth rates have been obtained with other species in captivity but

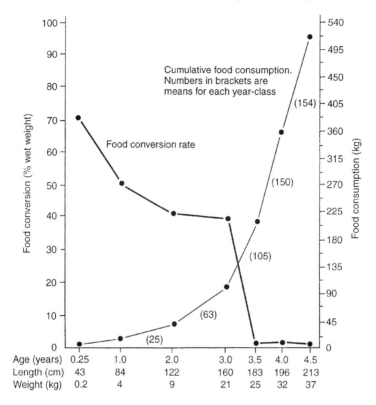

Figure 7.1 Growth and food consumption of American alligators in controlled environment chambers maintained at a constant 28°C for most of the growing period. (Redrawn from Joanen and McNease 1987.)

this early work with alligators demonstrated the level of success that could be expected with careful husbandry and reasonably sophisticated facilities. A variety of approaches to captive rearing and conservation are included in the following selection of national programmes. For convenience, the word 'farm' will be used for all crocodile rearing establishments.

7.4 MANAGEMENT PROGRAMMES

7.4.1 America

Years of protection enabled the American alligator to make a spectacular recovery in parts of its former range, notably in the state of Louisiana and parts of Florida, Texas and South Carolina. The species was transferred from Appendix I to Appendix II of CITES in 1979.

Experimental harvests conducted over a decade in Florida have indicated that an annual offtake of 13% of the estimated population of 'commercial sized' alligators ($\geqslant 122$ cm in length) has no measurable negative effect on alligator populations (Woodward *et al.* 1992). Similar studies on egg collection suggest that up to 50% of the annual production of eggs from relatively dense populations can be harvested sustainably (Wiley and Jennings 1990).

Each of the states mentioned above has an alligator utilization programme but more than 70% of production is from Louisiana, where management is controlled by the Louisiana Department of Wildlife and Fisheries (LDWF). Most of the details that follow are from Elsey *et al.* (1994).

(a) Farming/ranching

Eggs are collected in the wild (ranched), under licence and mostly from privately owned wetlands. About 100 'farms' were licensed in 1993 and in total they acquired more than 122 000 hatchlings from ranched eggs during that year.

It is mandatory in Louisiana for alligators up to 4 ft (1.2 m) long to be kept in heated enclosures, and farms are inspected by LDWF officers. The enclosure design favoured by most farmers consists of a large, insulated shed with a concrete floor in which hot-water pipes are embedded. A central aisle gives access on either side to a row of concrete tanks in which the alligators are kept. A development of the rearing shed arrangement, adopted to maximize the use of space and so reduce heating costs per alligator, makes use of fibreglass tanks stacked in tiers above concrete tanks at ground level. Servicing of the upper tiers has to be done with ladders. This less convenient arrangement has been adopted by about 25% of Louisiana's alligator farmers (Joanen and McNease 1990).

It is a condition of the ranching licence that a percentage of eggs hatched must be returned to the wild. The percentage is determined on a sliding scale according to average size at release such that, for example, at 3 ft long (91 cm) 29.6% of eggs hatched must be released; but at 4 ft long (122 cm) only 17% need be released. In 1993 a total of 28 512 alligators were released with an average length of 106.7 cm. Telemetry studies and recaptures from earlier releases indicate that survival and growth rates are satisfactory.

While most farm stock is acquired from the wild there is strong interest, in Louisiana as elsewhere, in captive breeding. However, despite many years of research at the Rockefeller Refuge, and on private farms, productivity from captive breeding remains disappointing (Elsey *et al.* 1994). Even when good rates of nesting have been achieved, egg fertility and hatching rates have usually been inferior to those achieved with eggs from the wild.

(b) Hunting harvest

A commercial harvest of mainly adult alligators is taken each September. In 1993 about 24 000 were killed by some 1600 licensed hunters throughout Louisiana.

Individual nuisance alligators, which have been the subject of complaints, are caught or killed by licensed trappers who are issued with tags in respect of each animal (820 in 1992).

(c) Sale of skins and other parts

Captive-reared alligators are slaughtered at about 100–120 cm long for maximum returns on feeding costs. Alligators killed in the wild by the annual hunt are mostly twice that length.

Raw salted skins are sold at auction and the value depends on length and quality (elsewhere in the world crocodilian skins are sized according to belly skin width rather than length). A grade 1 skin will be well preserved and free from any blemishes or cuts on the belly area. About 86% of Louisiana skins are exported to tanneries in Europe, Japan and Singapore. Alligator meat is sold to individuals, restaurants and fish markets both within Louisiana and interstate. Some teeth and skulls are sold to the jewellery trade and as biological specimens.

In 1990 it was estimated that the value of the entire alligator harvest to the state of Louisiana was about US$18 million per annum (Joanen 1990).

7.4.2 Africa

The Nile crocodile remains on Appendix I of CITES but several African countries have obtained downlisting to Appendix II. Zimbabwe led the way in this and obtained downlisting for the Nile crocodile in 1983. A few other African countries have obtained downlisting subject to an export quota. There are now more than 40 crocodile farms in Zimbabwe, over 30 in South Africa and smaller numbers of farms in several other African countries. Progress in southern Africa has been documented over many years (Pooley 1971; Blake and Loveridge 1975; Child 1987). To begin with, there were heavy losses of eggs and hatchlings and it was by no means certain that crocodile rearing could be commercially viable. Major constraints and advances can be summarized, mainly with reference to southern Africa, under the following headings.

(a) Egg collection and incubation

In Zimbabwe quotas of eggs are collected under permit from the Department of National Parks and Wildlife Management. Because egg-laying extends over a period of weeks it is not practicable to fulfil quotas by locating eggs soon after they are laid, so collections are made late in the season. At the rearing stations eggs are left to complete their incubation in boxes of vermiculite in rooms where temperature and humidity are kept within acceptable limits (28–34°C). Hatch rates of 80–95% are reported to be regularly achieved (Hutton and Van Jaarsveldt 1987). Early attempts at incubation by reburying eggs in protected patches of ground were less successful.

In Ethiopia, on islands of the Rift Valley Lakes, it has proved most successful to protect nests on communal nesting grounds by piling rocks on the sand above each nest. A watchman is then employed to camp at each nesting ground until the young can be heard calling – at which point the eggs are dug out and taken into captivity for rearing (Hailu 1990). In South Africa, harvesting from the wild has been illegal and farmers have been entirely dependent upon captive breeding for the supply of eggs, the original breeding stock having been imported from Zimbabwe and Botswana. A limited ranching scheme with egg collection from privately owned lands is now expected to develop.

(b) Rearing of hatchlings

In Africa crocodile farms have generally succeeded with hatchlings only after some form of heated enclosure has been installed. Water in small ponds does not retain its heat for very long after sunset and it can quickly become too hot on a sunny day. Overheating outdoors can easily be prevented by providing shade but cold presents a much more expensive problem. Cold reduces appetite, slows digestion and, as a stress factor (singly or in combination with others such as handling or overcrowding), can be expected to suppress the immune system so that hatchlings become more susceptible to disease (Smith and Marais 1994; Turton *et al.* 1994).

In some cases, original enclosures had to be modified or rebuilt and a variety of pen designs and heating methods are now used. Farms are showing very different results; in South Africa a growth period of 24–30 months appears to be the norm for crocodiles to reach a slaughter length of about 135 cm (Smith and Marais 1990).

(c) Juvenile crocodiles

Growth rates vary considerably between individuals but at the age of a few months, when captives are approaching 1 metre in length, they are generally tougher and can survive in outdoor pens. Growth will be slower but some farmers, even in relatively cool South Africa, choose to save on heating costs by transferring stock to open-air enclosures at the age of 9–10 months (Smith and Marais 1990). In Zimbabwe there is legal provision, under the ranching scheme, for returning up to 5% of hatchlings to the wild when they reach 120–150 cm in length. This provision was very lightly used through the 1980s as wild populations were considered to be adequate in areas of suitable habitat (Child 1987). More recently, releases of 2% have been made against heavier egg collecting but the survival rate has been disappointing. Many released animals are drowned in fishing nets (D. Hearh, pers. comm.).

Once young crocodiles are past the delicate stage the problem of food supply becomes increasingly important and is probably a limiting factor for most farms. Smith and Marais (1990) reported that in South Africa supplies of protein (mainly offal from the poultry industry) were becoming increasingly difficult to obtain and some farmers were having to send vehicles nearly 200 km to collect even moderate loads. In Zimbabwe farmers have depended upon fish supplemented by game meat from controlled hunting programmes (Hutton and Van Jaarsveldt 1987) but supplies of animal protein, in southern Africa and elsewhere, are increasingly in demand for more direct human needs.

Most skins from African crocodile farms are exported but a few are tanned locally and manufactured into articles for sale to tourists where retail outlets exist.

7.4.3 Papua New Guinea

(a) Background
Ninety-five per cent of the land of Papua New Guinea (PNG) is held under a system of traditional tenure, and wildlife belongs to the families and clans who own the land. Two crocodile species, *C. porosus* and the endemic freshwater crocodile *C. novaeguineae* are widely distributed within PNG and both species are listed on Appendix II of CITES.

Commercial hunting only began in the 1930s, with heavy exploitation dating from the early 1950s. Nearly 350 000 skins were exported between 1961 and 1966 (Hollands 1987). Hunting is mainly conducted at night by villagers paddling dugout canoes and finding crocodile eye reflections with a torch (Plate 1). Small crocodiles are usually killed with a spear but a shotgun may be used for bigger ones.

By the late 1960s the consequences of over-exploitation were becoming apparent; by the mid 1970s the mean belly width of *C. porosus* skins had fallen to about 18 cm, indicating that the average animal being killed was less than 1 m long. Older animals were becoming scarce.

(b) A policy for the crocodile industry
A national policy for restructuring the industry was formulated at the end of the 1960s (Downes 1971). The policy had a social element that reflected official concern with rural economies and the problem of urban drift. Special emphasis was placed on involvement and benefit at village level. Villagers were persuaded to accept a ban on commerce in skins of more than 51 cm belly width in order to protect breeding animals (killing for meat or other traditional uses was not affected). It was to be another decade before a ban on the wasteful killing of very small animals (< 18 cm belly width) could be introduced. These small crocodiles became the basis of a crocodile ranching scheme.

In accordance with national policy, crocodile ranching in PNG was originally envisaged as a cottage industry with crocodiles being caught as hatchlings and reared to commercial size in numerous village farms. Egg collection was discouraged because of the additional problems of collection, transportation and incubation of eggs under very difficult conditions. In 1977 a technical and economic assistance project became operational. Funded by the United Nations Development Programme (UNDP) and executed by the UN Food and Agriculture Organization (FAO), the project ran for approximately 5 years (FAO 1981).

(c) Problems at the village level

Despite the best efforts and inputs by government and the assisting project, the original scheme could not be made to succeed. The various difficulties have been described (Bolton and Laufa 1982; Hollands 1987) and among them were:

- seasonal fluctuations of water level amounting to several metres in some regions
- seasonal difficulties in obtaining fish, and the impossibility of storing fish, in remote settlements
- the problem of cold in small outdoor pools
- inadequate care of crocodiles when owners were away

Only about 15% of village holdings were considered to be reasonably successful for more than one season.

(d) Change of priorities

It had been foreseen that large, technically sophisticated farms could eventually be useful but with the emphasis at village level such farms had not been encouraged by government. It now became necessary to give them high priority so that villagers could sell young live crocodiles instead of trying to rear them for long periods in the villages.

Two farms were set up by the private sector in association with modern poultry farms; poultry offal providing an assured supply of fresh food. On this diet *Crocodylus porosus* reach culling size in 2½ years (D. Wilken, pers. comm.) although in feeding trials fish-fed crocodiles gained weight rather faster (Bolton 1989). On a fish diet, *C. porosus* reached an average length of 1.8 m in 3 years. Living wild in the comparable climate of Northern Australia, the same species takes more than 6 years to reach 1.8 m (Webb *et al.* 1978).

Purchasing patrols were established to buy young crocodiles from villagers and transfer them to government collection centres until enough had been assembled to cover the costs of transportation to a commercial farm. Villagers were paid well for healthy crocodiles and it was this which convinced them that spearing small animals for their skins was in nobody's best interests. By the end of 1981 there were some 30 000 crocodiles in captivity, more than half of them in the two largest farms (Plate 2). Even then, however, mortality among purchased crocodiles of less than about 50 cm in length remained disappointingly high.

A wild crocodile monitoring programme was established in an attempt to detect and follow trends in exploited wild populations. The vast swamplands of PNG, with dense, floating rafts of vegetation and choked waterways, present exceptional difficulties for crocodile census work. The method adopted was based on a system of nest counting along survey routes flown annually by helicopter (Graham 1980; Hollands 1984). A nesting index was derived for each species in the area surveyed, although nest numbers are not a true measure of breeding population size because not all adult females try to breed in any one year. Useful information is also provided by the composition (skin sizes and species) of the annual harvest as revealed by the export documentation.

(e) The existing situation
The crocodile industry of PNG is no longer in decline. By 1987 the partial shift to ranching, together with the size limitations on saleable skins and a general improvement in skinning and preservation standards, had brought about a 43% increase in production from a reduced offtake (Hollands 1987).

But the ranching component of PNG's crocodile industry has not reached its full potential. In 1994 there were 38 000 crocodiles in six farms, with over 32 000 of them in one of the two, original, commercial establishments (Solmu 1994). Existing food supplies, available for crocodiles, could support more than twice that number. Acquisition of stock is a major constraint; it is simply uneconomic for the private sector to maintain buying patrols in the most remote regions where travel can involve days of difficult progress through densely vegetated swampland. Moreover, mortality of very young crocodiles has remained unacceptably high so that commercial buyers are reluctant to purchase animals of less than about 56 cm in length. At one farm new intakes are kept in heated quarantine pens and given nutritional drenches and routine medication but there is still about 6% mortality compared with less than 1% for captive-bred crocodiles of comparable size (D. Wilken, pers. comm.).

Hunted skins, within the 18–51 cm size range, still make up the bulk of annual exports (Figure 7.2) but the monitoring programme indicates that wild crocodile populations are not diminishing as a result of hunting and ranching. The annual nest counts have shown numbers of *C. porosus* nests to fluctuate above the 1981 index. Numbers of *C. novaeguineae* nests have fluctuated more closely around the 1981 level (Solmu 1994). The proportion of animals which build nests could be associated with factors such as seasonal water levels and the extensive fires which destroy cover during a prolonged dry season. The spread of cultivation along river banks which

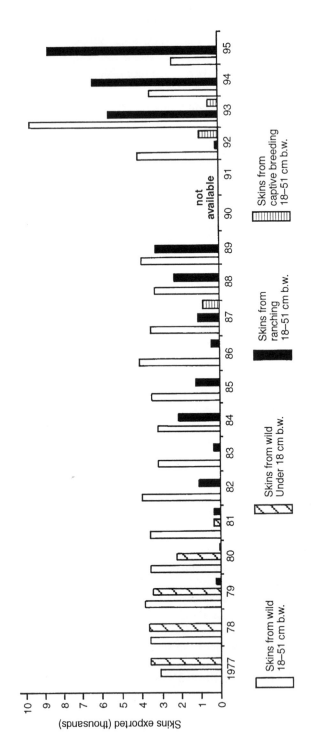

Figure 7.2 Skins of *Crocodylus porosus* exported from Papua New Guinea from 1977 to 1995. (Source: Solmu 1994 and personal correspondence.)

were formerly used by nesting crocodiles has yet to be quantified but it appears to be accelerating (D. Wilken, pers. comm.).

7.4.4 India

All three of India's crocodilians were reduced to remnant populations by the end of the 1960s. The mugger (*C. palustris*) was still widely distributed but populations were small and fragmented. The saltwater crocodile (*C. porosus*) occurred only on a small part of the east coast and on the Andaman and Nicobar Islands in the Indian Ocean. The gharial (*Gavialis gangeticus*) had been reduced to about 100 individuals in India with a total world population of about 150 (Whitaker 1987).

Management policy was directed towards restoring populations to more viable levels and FAO provided technical and financial assistance with a long-term view of resource development (Bustard 1974). Priority was given to the gharial but a programme for all three species involved egg collection followed by incubation, captive rearing and release into selected sanctuaries. These measures may well have saved the gharial from total extinction; there are now believed to be at least 1500 gharial surviving in the wild, though more than 3300 have been released (Rao and Singh 1994) and a large proportion of the survivors are in a single sanctuary – the National Chambal. Muggers are still widely distributed with perhaps 3000–5000 in the wild and 12 000 in captivity. Saltwater crocodiles in the wild, however, remain confined to only two localities on the Indian mainland where not more than about 1000 exist in total. In India there are less than 1000 in captivity (Whitaker 1990; Anon. 1993).

Captive breeding is now straining the resources of the existing centres and further releases are considered to be of doubtful value because of growing public resistance to the presence of wild crocodiles in areas of human habitation. Official policy is to discontinue captive breeding and to maintain protection of crocodilians in the existing sanctuaries (Whitaker 1990; Ross 1994). No utilization programme has been authorized and it remains to be seen how effectively India's crocodilians can be protected without commercial incentives.

7.5 TRENDS AND UNCERTAINTIES

7.5.1 Ranching, farming and conservation

Ranching crocodilians has proved to be more profitable, with present levels of technology, than has closed-cycle farming. Indeed, most closed-cycle farms are heavily dependent on income from tourism. In some countries, such as China, ranching is no longer an option because destruction of both the animals and their habitat has gone too far. In other parts of the world ranching can be practised only on a very small scale, and where, as in South Africa, the main constraint is lack of suitable habitat, it is unlikely that prospects for ranching will improve in the foreseeable future.

Even where potentially suitable habitat remains, species perceived as being dangerous to humans or livestock may be regarded as an unnecessary hazard once people have experienced life without them. In western societies the removal of nuisance crocodilians, including American alligators, has commonly been the first component of an active management policy – even though the American alligator does not often attack humans.

Nor can people in developing countries be expected to remain ever-tolerant of crocodiles. In Kenya, for example, people are increasingly coming into conflict with crocodiles, and attitudes are reported to be hardening (Soorae 1994). In the Solomon Islands, crocodiles (*C. porosus*) are still widespread but very scarce and many riverside communities are hopeful of exterminating them completely (Bolton 1988; Messel and King 1990). It may not be possible to restore crocodile numbers to the point where they could sustain a widespread ranching programme in places such as the Solomon Islands. A small, cautious, ranching, release and restoration scheme, based on such isolated sanctuaries as might be immediately secured and locally managed, could have more chance of success, but the opportunity for extensive ranching has probably passed.

Closed-cycle farms are not entirely without conservation value; they offer a number of possible benefits, as do zoos (Chapter 13). These 'zoo benefits' are being realized in different parts of the world by farms, with and without a ranching component, and at least no negative effects are apparent.

7.5.2 The commercial factor

In late 1990, as a result of a global economic downturn, there was a sudden drop in demand for crocodile skins. Prices fell and there was a predictable range of responses from producers; ranching quotas were not taken up, skins were stockpiled by those who could afford it, and one European tannery closed temporarily. Trade slowly began to recover but the market crash demonstrated the vulnerability of farming and ranching schemes with substantial capital investment and high running costs. In contrast, skin-hunters, legal and otherwise, can suspend their activities comparatively easily.

It has been pointed out that regulation of international trade is a primary mechanism for controlling use and ensuring sustainability (Ross and Godshalk 1995). But international commodity prices are notoriously difficult to regulate. Future trade in crocodilian skins could easily become less stable as increasing numbers of captive-bred caimans from South and Central America are added to the wild harvest, while at the same time the global focus of tanning and manufacture shifts eastwards to Singapore, China and Taiwan.

There is still substantial slack in this new industry and no doubt greater efficiencies will be made at all levels; cooperative marketing, for example, has already proved its worth (King and Ross 1991). Rearing costs, however, could prove very difficult to reduce, dependent as they are upon the price of animal protein. Selective breeding of captive stock may select for faster-growing strains of crocodilians but there is also the possibility that, at any time, market forces will bring about a demand for skins of larger sizes – extending the rearing period into that part of the animal's growth curve where diminishing returns begin to take effect.

Research into crocodilian nutrition has shown that it is possible to rear these animals entirely on commercially available dry pelletized rations containing a minimum of 45% protein (Brisbin *et al.* 1990; Staton *et al.* 1990). Dry pellets are easy to handle and cheap to store but, in practice, the economics of using this type of feed, as with all manufactured animal feeds and supplements, can be expected to depend very much upon the farmer's geographical location and particular circumstances.

7.6 CONCLUSION

For a few species of crocodilians, the conservation measures adopted during the past three decades have been highly successful in diminishing the impact of deliberate persecution. Populations of these crocodilians have now been brought to the point where, in common with most other wildlife, the main constraint on their numbers is shortage of habitat rather than hunting.

Part of the rationale of sustained yield harvesting of crocodilians is the belief, or at least the hope, that it will provide an incentive for the protection of habitat of the valued species (Thorbjarnarson 1992; Ross and Godshalk 1995). But examples of land-use decisions having been made in favour of crocodiles are less easy to find than are the more technical advances outlined above. In some localities where a site such as an island or part of a lake is involved, as in the Rift Valley of Ethiopia, crocodile ranching has taken precedence over fishing interests to the extent that crocodile nesting grounds have been protected from disturbance, but even here there is no real conflict over the future of the habitat; nobody has yet proposed that the sandbanks be converted to anything else. In Papua New Guinea it is too early to say whether the value of crocodiles will be significant in protecting nesting grounds from clearance and cultivation in areas of potential conflict.

Crocodilians have proved to be particularly suitable animals for sustained yield harvesting, and commercial incentives have been applied very successfully to their conservation. But the crocodilian case study also demonstrates some of the limitations, as well as the strengths, of utilization as a conservation tool.

ACKNOWLEDGEMENTS

Thanks to Perran Ross, David Wilken, Godfrid Solmu and Lorraine Collins for helpful information.

REFERENCES

Anon. (1993) Information from Indian sources reported in *Crocodile Specialist Group Newsletter*, IUCN, **12** (1), 10.
Blake, D.K. and Loveridge, J.P. (1975) The role of commercial crocodile farming in crocodile conservation. *Biol. Conserv.*, **8**, 261–72.
Bolton, M. (1988) Feasibility study of crocodile farming in the Solomon Islands. FAO Field Document TCP/SOI/6753, FAO, Rome.
Bolton, M. (1989) *The Management of Crocodiles in Captivity*. FAO Conservation Guide No. 22, FAO, Rome.
Bolton, M. and Laufa, M. (1982) The crocodile project in Papua New Guinea. *Biol. Conserv.*, **22**, 169–79.
Brisbin, I.L., Standora, E.A. and Vargo, M.J. (1982) Body temperature and behaviour in American alligators during cold winter weather. *Am. Midl. Nat.*, **107**, 209–18.
Brisbin, I.L., McCready, C.D., Zippler, H.S. and Staton, M.A. (1990) Extended maintenance of American alligators on a dry formulated ration, in *Crocodiles. Proc. 10th Working Meeting of the IUCN Crocodile Specialist Group*, IUCN, Gland, Switzerland, pp. 16–31.
Bustard, H.R. (1974) India. A preliminary survey of the prospects for crocodile farming. FAO Field Document FAO/IND/71/033, FAO, Rome.
Chabreck, R.H. and Joanen, T. (1979) Growth rates of American alligators in Louisiana, *Herpetologica*, **35**, 51–7.
Child, G. (1987) The management of crocodiles in Zimbabwe, in *Wildlife Management; Crocodiles and Alligators* (eds G.J.W. Webb, S.C. Manolis and P.J. Whitehead). Surrey Beatty, Chipping Norton, Australia, pp. 49–62.
Coulson, T.D., Coulson, R.A. and Hernandez, T. (1973) Some observations on the growth of captive alligators, *Zoologica* (New York), **58**, 45–52.
Downes, M.C. (1971) Regional situation report: Papua New Guinea, in *Crocodiles*, IUCN Publ. New Series Supp. Paper No. 32, pp. 41–3.
Elsey, R., Joanen, T. and McNease, L. (1994) Louisiana's alligator research and management programme: an update, in *Crocodiles. Proc. 12th Working Meeting of the IUCN Crocodile Specialist Group*, IUCN, Gland, Switzerland, pp. 199–229.
FAO (1981) Assistance to the Crocodile Skin Industry, Papua New Guinea: Project Findings and Recommendations, PNG/74/029, Terminal Report, FAO, Rome.
Ferguson, M.W.J. (1985) Reproductive biology and embryology of the crocodilians, in *Biology of the Reptilia*, Vol. 14 (eds C. Gans, F.S. Billett and P.F.A. Maderson). John Wiley and Sons, New York, pp. 329–491.
Ferguson, M.W.J. and Joanen, T. (1982) Temperature of egg incubation determines sex in *Alligator mississippiensis. Nature, Lond.*, **296**, 850–3.

Fuchs, K.H.P. (1975) *The Chemistry and Technology of Novelty Leathers*. FAO, Rome.

Graham, A.D. (1980) Monitoring workplan, FAO/UNDP PNG/74/029. Assistance to the Crocodile Skin Industry project document, Wildlife Division, Port Moresby, Papua New Guinea.

Hailu, T. (1990) The method of crocodile hatching adopted at Arba Minch crocodile farm, Ethiopia, in *Crocodiles. Proc. 10th Working Meeting of the IUCN Crocodile Specialist Group*, IUCN, Gland, Switzerland, pp. 173–9.

Hemley, G. and Caldwell, J. (1984) The crocodile skin trade since 1979, in *Crocodiles, Proc. 7th Working Meeting of the IUCN Crocodile Specialist Group*, IUCN, Gland, Switzerland, pp. 398–412.

Hollands, M. (1984) A preliminary examination of crocodile population trends in Papua New Guinea 1981–1984. Report to the 7th Working Meeting of the IUCN Crocodile Specialist Group, Caracas, Venezuela, October 1984.

Hollands, M. (1987) Management of crocodiles in Papua New Guinea, in *Wildlife Management; Crocodiles and Alligators* (eds G.J.W. Webb, S.C. Manolis and P.J. Whitehead). Surrey Beatty, Chipping Norton, Australia, pp. 73–89.

Hutton, J.M. and Van Jaarsveldt, K.R. (1987) Crocodile farming in Zimbabwe, in *Wildlife Management; Crocodiles and Alligators* (eds G.J.W. Webb, S.C. Manolis and P.J. Whitehead). Surrey Beatty, Chipping Norton, Australia, pp. 323–7.

Inskipp, T. and Wells, S. (1979) *International Trade in Wildlife*. Earthscan Publications, IIED, London.

Joanen, T. (1990) Alligator farm production in the United States, in *Crocodiles. Proc. 10th Working Meeting of the IUCN Crocodile Specialist Group*, IUCN, Gland, Switzerland, p. 317.

Joanen, T. and McNease, L. (1975) Notes on the reproductive biology and captive propagation of the American alligator. *Proc. Ann. Conf. Southeastern Assoc. Game and Fish Comm.*, **29**, 407–15.

Joanen, T. and McNease, L. (1976) Culture of immature American alligators in controlled environmental chambers. *Proc. World Mariculture Society*, 7, 201–11.

Joanen, T. and McNease, L. (1979) Culture of the American alligator. *International Zoo Yearbook*, **19**, 61–6.

Joanen, T. and McNease, L. (1987) Alligator farming research in Louisiana, USA, in *Wildlife Management; Crocodiles and Alligators* (eds G.J.W. Webb, S.C. Manolis and P.J. Whitehead). Surrey Beatty, Chipping Norton, Australia, pp. 329–40.

Joanen, T. and McNease, L. (1990) Alligator farm design in Louisiana, in *Crocodiles. Proc. 10th Working Meeting of the IUCN Crocodile Specialist Group*, IUCN, Gland, Switzerland, pp. 268–74.

King, F.W. and Ross, J.P. (1991) Editorial. *Crocodile Specialist Group Newsletter*, IUCN, 10 (1), 2.

Lang, J.W. (1987) Crocodilian thermal selection, in *Wildlife Management; Crocodiles and Alligators* (eds G.J.W. Webb, S.C. Manolis and P.J. Whitehead). Surrey Beatty, Chipping Norton, Australia, pp. 301–17.

Lang, J.W. and Andrews, H.V. (1994) Temperature-dependent sex determination in crocodilians. *Journal of Experimental Zoology*, 270, 28–44.

Messel, H. and King, F.W. (1990) The status of *Crocodylus porosus* in the Solomon

Islands, in *Crocodiles. Proc. 10th Working Meeting of the IUCN Crocodile Specialist Group*, IUCN, Gland, Switzerland, pp. 39–69.

Pooley, A.C. (1971) Crocodile rearing and restocking, in *Crocodiles*, IUCN Publ. New Series Supp. Paper No. 32, pp. 104–30.

Rao, R.J. and Singh, L.A.K. (1994) Status and conservation of the gharial in India, in *Crocodiles. Proc. 12th Working Meeting of the IUCN Crocodile Specialist Group*, IUCN, Gland, Switzerland, pp. 84–97.

Ross, J.P. (1994) Reported information from D. Jelden and R. Whitaker. *Crocodile Specialist Group Newsletter*, IUCN, **13** (4), 10.

Ross, J.P. and Godshalk, R. (1995) Sustainable use, an incentive for the conservation of crocodilians. Paper presented at the 2nd International Conference on Wildlife Management in Amazonia, May 1995, Iquitos, Peru.

Smith, G.A. and Marais, J. (1990) Crocodile farming in South Africa: the impact of farming technology on production efficiency, in *Crocodiles. Proc. 10th Working Meeting of the IUCN Crocodile Specialist Group*, IUCN, Gland, Switzerland, pp. 201–15.

Smith, G.A. and Marais, J. (1994) Stress in crocodilians: the impact of nutrition, in *Crocodiles. Proc. 12th Working Meeting of the IUCN Crocodile Specialist Group*, IUCN, Gland, Switzerland, pp. 2–38.

Solmu, G.C. (1994) Status of *Crocodylus porosus* and *Crocodylus novaeguineae* populations in Papua New Guinea, 1981–1994, in *Crocodiles. Proc. 12th Working Meeting of the IUCN Crocodile Specialist Group*, IUCN, Gland, Switzerland, pp. 77–102.

Soorae, P.S. (1994) Nile crocodile in Kenya. *Crocodile Specialist Group Newsletter*, IUCN, **13** (3), 5.

Staton, M.A., McNease, L., Theriot, L. and Joanen, T. (1990) Pelletized alligator feed: an update, in *Crocodiles. Proc. 10th Working Meeting of the IUCN Crocodile Specialist Group*, IUCN, Gland, Switzerland, pp. 216–21.

Thorbjarnarson, J. (1992) *Crocodiles: an Action Plan for their Conservation*, IUCN Crocodile Specialist Group Action Plan, IUCN, Gland, Switzerland.

Turton, J.A., Ladds, P.W., Manolis, S.C. and Webb, G. (1994) The influence of water temperature and clutch of origin on stress in farmed *Crocodylus porosus* hatchlings, in *Crocodiles. Proc. 12th Working Meeting of the IUCN Crocodile Specialist Group*, IUCN, Gland, Switzerland, p. 64.

Webb, G.J.W., Messel, H., Crawford, J. and Yerbury, M.J. (1978) Growth rates of *Crocodylus porosus* (Reptilia: Crocodilia) from Arnhem Land, Northern Australia. *Australian Wildlife Research*, **5**, 385–99.

Webb, G.J.W., Manolis, S.C., Dempsey, K.E. and Whitehead, P.J. (1987) Crocodilian eggs: a functional overview, in *Wildlife Management; Crocodiles and Alligators* (eds G.J.W. Webb, S.C. Manolis and P.J. Whitehead). Surrey Beatty, Chipping Norton, Australia, pp. 417–22.

Whitaker, R. (1987) The management of crocodilians in India, in *Wildlife Management; Crocodiles and Alligators* (eds G.J.W. Webb, S.C. Manolis and P.J. Whitehead). Surrey Beatty, Chipping Norton, Australia, pp. 64–72.

Whitaker, R. (1990) Summary report from the western Asia region. *Crocodile Specialist Group Newsletter*, IUCN, **9** (January–March), 6.

Wilbur, C.G. (1960) Effect of temperature on the heart in the alligator. *Am. J. Physiol.*, **198**, 861–3.

Wiley, E.N. and Jennings, M.L. (1990) An overview of alligator management in Florida, in *Crocodiles. Proc. 10th Working Meeting of the IUCN Crocodile Specialist Group*, IUCN, Gland, Switzerland, pp. 274–85.

Woodward, A.R., Moore, C.T. and Delaney, M.F. (1992) Experimental Alligator Harvest, Final Report. Study No. 7567, Florida Game and Freshwater Fish Commission, Gainesville, USA.

8

Gamebirds: management of the grey partridge in Britain

Nicholas J. Aebischer

8.1 INTRODUCTION AND HISTORICAL OVERVIEW

The grey partridge *Perdix perdix* was originally a bird of the temperate steppe grasslands of central Europe and Asia. It adapted readily to an open arable landscape and vastly expanded its range as agricultural development spread westward across Europe over the last eight millennia. At its maximum extent, this species occurred throughout Europe and north-west Asia east to Mongolia. The range was further extended by successful introductions into North America at the beginning of the twentieth century; it is now established in the northern half of the United States and in south-west Canada.

In the UK, a combination of land enclosure, increased cultivation and intensive predator control in the eighteenth and especially nineteenth centuries boosted numbers of grey partridges considerably (Potts 1986). At the same time, sporting interest in the species developed, and it was the most popular gamebird of the last century. Bag records show that around two million grey partridges were killed annually on a sustainable basis between 1870 and 1930 (Tapper 1992).

The same bag records indicate that after the Second World War, numbers of grey partridges fell dramatically in nearly all countries in which it was found (Potts 1986). The declines were accompanied by a retraction in range from the periphery inwards, and away from mountainous areas. Nowadays, grey partridges have vanished from Norway and, in Finland and Sweden, are restricted to the south. They have disappeared from Portugal and much of north-west Spain, southern Italy and southern Greece. In Ireland and Switzerland, they are almost extinct (Aebischer and Potts 1994).

In the UK, the average bag declined by over 80% between 1940 and

Conservation and the Use of Wildlife Resources. Edited by M. Bolton.
Published in 1997 by Chapman & Hall. ISBN 0 412 71350 0.

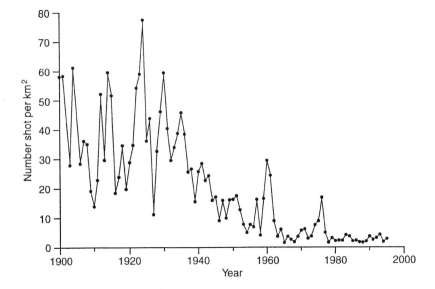

Figure 8.1   Mean annual number of grey partridges shot per km^2 from 12 estates in southern and eastern England for which records extend back to 1900. After Tapper (1992) updated.

1990 (Figure 8.1). By the mid 1960s, the situation aroused such concern in the shooting world that The Game Conservancy Trust launched a research project in 1968 to investigate the causes of the decline. The research was carried out on 62 km^2 of the Sussex Downs, in southern England, and much of what is currently known about the ecology and population dynamics of the grey partridge in Britain, described in the next section, stems from that project (Potts 1980, 1986). Although detailed work within the study area ceased in 1984, the project continues to this day with ongoing monitoring of grey partridge demography by means of annual counts, and monitoring of summer insect and weed abundance in the bird's arable environment (Aebischer 1991a; Potts and Aebischer 1995).

The monitoring shows that the density of spring pairs in Sussex continued to fall after the start of the study, from an initial 20 pairs/km^2 to under 5 pairs/km^2 in recent years (Figure 8.2a). The trend matches the national picture as given by the Common Bird Census of the British Trust for Ornithology, whose index of abundance of grey partridges on farmland dropped by 90% between its start in 1962 and 1994 (Figure 8.2b). The bird is currently listed as a UK Red Data species (Batten *et al.* 1990).

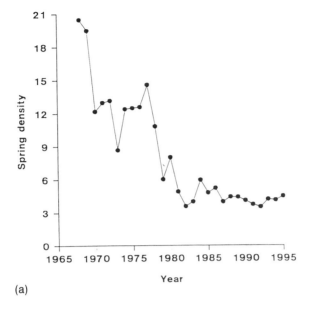

(a)

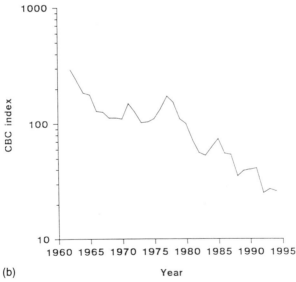

(b)

Figure 8.2 Changes in the annual abundance of grey partridges in the UK: (a) spring density (pairs/km^2) on The Game Conservancy Trust's main study area in Sussex, 1968–1993, from Potts and Aebischer (1995) updated; (b) national index of abundance based on the Common Bird Census of the British Trust for Ornithology, 1962–1994, from Marchant *et al.* (1990) updated.

8.2 LIFE CYCLE AND ECOLOGY

The life cycle of the grey partridge is summarized in Figure 8.3. The species is monogamous and, after pair formation in the spring, the female seeks out a suitable piece of cover in which to nest. The nest is constructed on the ground, and is typically a scrape lined with grasses. Nest sites are chosen so as to be free from water seepage and local flooding during incubation. The nest is usually well hidden by being built in rank dead grass or other tall vegetation, such as that found on uncut field margins, hedge bottoms or even autumn-sown cereals. The average clutch size is 15 eggs, the highest of all known bird species. Incubation lasts 23–25 days, and is by the female alone. During this period, there is a high risk of predation by avian and ground predators. The main egg predators are carrion crows *Corvus corone* and magpies *Pica pica*, as well as rats *Rattus rattus*, hedgehogs *Erinaceus europaeus* and mustelids, while the sitting hen may be killed by foxes *Vulpes vulpes*, cats *Felis catus* or stoats *Mustela erminea*. A female whose clutch is lost will usually lay a replacement clutch. The Sussex study found that the rate of nest predation was density-dependent, and that the slope of the density-dependent relationship was steepest in the absence of predation control (Figure 8.4).

Within a few hours of hatching, the chicks are mobile and are led away from the nest by the parents. They feed themselves, but require brooding by the adults to keep warm in cold weather. During the first 2 weeks after

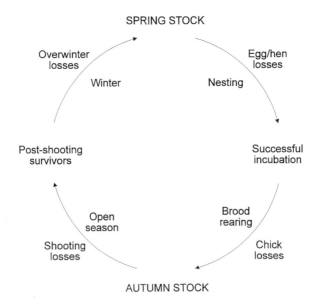

Figure 8.3 Schematic representation of the life cycle of the grey partridge.

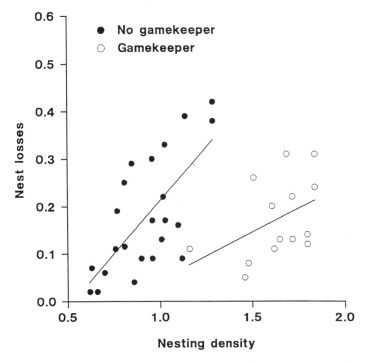

Figure 8.4 Density-dependent relationships between the annual proportion of grey partridge nests lost to predation ($-\log_{10}$) and the annual density of nesting females per km^2 ($\log_{10}$), for years with and those without a gamekeeper (from Potts 1980).

hatching, the chick diet comprises almost entirely of insects, a rich source of protein (Ford *et al.* 1938). Laboratory experiments have shown that feather development is much faster for chicks on a high-protein diet than on a low-protein one (Potts 1986). This is important because, by the age of 10 days, the chicks are able to flutter away from danger, thereby increasing their chances of escaping predation. After the first fortnight, the proportion of weed seeds and other vegetable material in the diet gradually increases to match that in the adult diet. By the age of 12 weeks, the chicks are fully grown and begin to moult into adult plumage.

Most chick mortality in the wild occurs during the first few weeks after hatching, and two sources of field data support the link with insect abundance: (1) annual chick survival rates from the Sussex study are positively correlated with an index of insect abundance in cereals obtained in the summer (Figure 8.5a), and (2) the survival of individual broods followed by radio-tracking is positively correlated with the proportion of certain preferred insect taxa in the diet, based on faecal analysis (Figure 8.5b). Annual

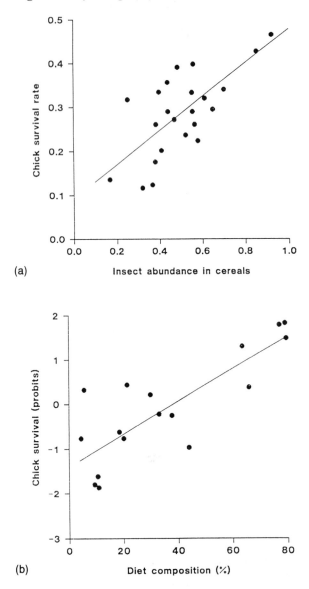

Figure 8.5 Relationships between grey partridge chick survival (to 6 weeks) and chick-food insects: (a) annual survival rates in Sussex in relation to abundance indices of chick-food insects sampled in cereals in mid-June (time of peak chick hatch), 1970–1992; (b) survival of chicks in 17 radio-tracked broods in relation to diet composition (percentage of caterpillars Symphyta and Lepidoptera, leaf beetles Chrysomelidae and weevils Curculionidae) determined by faecal analysis (source: Potts and Aebischer 1991; within-brood survival rates transformed to normal equivalent deviates (probits) to satisfy variance assumptions of regression).

chick survival rates are density-independent, and their high variability results from a combination of insect availability and weather. The radio-tracking further showed that grey partridge broods spent 97% of their time inside cereal crops, close to the field margin (Green 1984).

During the autumn and winter, grey partridges band together in small groups called coveys, often made up of a pair with their offspring together with odd 'barren' pairs or single adults. They are sedentary throughout western Europe, but migration occurs further east, where winters are particularly harsh. The coveys break up in early spring as pairing takes place. The pairs defend territories and space themselves out in relation to the availability of suitable nesting habitat. Overwinter losses average around 40%, which represents a combination of mortality and emigration (Potts 1980). The magnitude of these losses is dependent on the autumn density corrected for availability of nesting habitat (Figure 8.6).

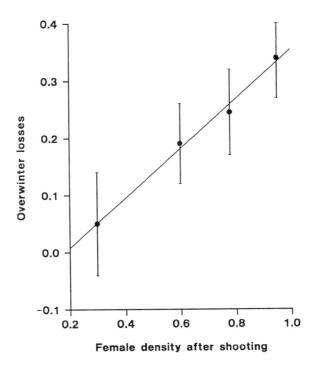

Figure 8.6 Density-dependent relationship between annual overwinter loss rate ($-\log_{10}$) of grey partridges in Sussex (mortality and emigration), and annual post-shooting density expressed as females per km of linear boundary per km^2 ($\log_{10}$) (source: Potts 1986).

8.3 CAUSES OF THE DECLINE

The decline of the grey partridge in Britain was the result of an interplay of factors whose origins lie in the modernization of agriculture after the Second World War (Potts 1980, 1986). It began with the introduction of herbicides in the early 1950s. Their use on cereals increased rapidly, with over 50% of cereal fields being treated by 1960, and close to 100% by 1965. This disrupted the food chain within the crops, as many of the partridge chick-food insects lived in and fed on the weedy understorey that herbicides destroyed. It is thought that herbicides were responsible for a 50% drop in the abundance of invertebrates within cereals (Southwood and Cross 1969). This matches the long-term trend in annual survival rates of partridge chicks (Figure 8.7), which averaged over 40% in the early 1950s, but dropped below 30% in the early 1960s.

Population modelling has shown that in itself, the fall in average annual

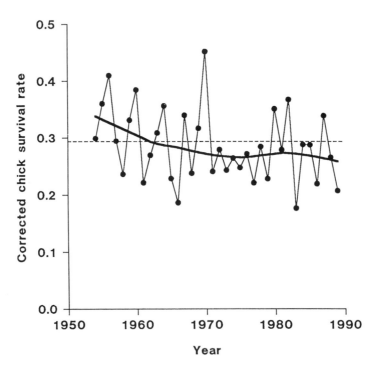

Figure 8.7 Annual chick survival rate (to 6 weeks) of grey partridges in Sussex, 1953–1990, corrected for weather and releasing of hand-reared birds. The thick line indicates the long-term trend; the dotted horizontal line is the minimum chick survival rate required to maintain the population.

chick survival rates would have occasioned only a relatively small drop in spring stocks because it was partially compensated by lower overwinter losses (Potts and Aebischer 1995). Its impact, however, was on autumn stocks and hence on the bag and the associated economic revenue from shooting. Many gamekeepers either lost their jobs or turned towards the rearing of pheasants *Phasianus colchicus*. As a result, the reduction in numbers of predators that was a traditional role of the gamekeeper ceased or became less intensive. This led to an increase in predation during the nesting season, i.e. higher female mortality and egg losses. The situation was exacerbated by the removal of nesting cover, as greater mechanization on farms led to the enlargement of fields by eliminating hedgerows and field boundaries. Further destruction of nesting cover resulted from the use of herbicides on hedge bottoms and field margins to prevent invasion of the crop by pernicious weeds on its periphery.

Thus the causes of the decline of the grey partridge in Britain, and probably over much of its range, were threefold (the 'three-legged stool'): a fall in the abundance of cereal insects, an increase in predation pressure during nesting, and a reduction in nesting cover. In retrospect, traditional management of grey partridge stocks for shooting before the Second World War sought to counteract the last two causes (the first cause was not an issue in the pre-intensification era). Modern management must take all three causes into account.

8.4   HABITAT MANAGEMENT

Putting the clock back in terms of agricultural change is not a feasible option. Habitat management for grey partridges at the end of the twentieth century must therefore be compatible with the requirements of modern agriculture. Several such methods are presented below.

8.4.1   Management of brood-rearing habitat

To tackle the dearth of chick-food insects in cereal crops that leads to low chick survival rates, a tried and tested method is the use of 'conservation headlands' (Sotherton 1991). This involves selective herbicide use on the outer 6 m of a cereal crop (crop margins are known as 'headlands'), according to prescriptions that are regularly updated to take into account the availability of new selective compounds. The aim is to encourage the development of annual arable weeds at ground level within the part of the crop most frequented by partridge broods, while preventing infestation by agriculturally pernicious and unacceptable weeds such as barren brome *Bromus sterilis* and cleavers *Galium aparine*. Thus non-residual herbicides that are specific to grasses or cleavers are allowed, as are fungicides (except the insecticidal pyrazophos) and insecticides up to 15 March (to enable the

Table 8.1 Percentage of grey partridge chicks that survived the first 6 weeks after hatching, in relation to the management of cereal headlands (outer 6 m of the crop) on experimental farms in Norfolk, 1984–1991. The headlands on one half of each farm were selectively sprayed (Conservation Headlands), and were conventionally farmed (fully sprayed) on the other half. After Sotherton *et al.* (1993)

	1984	1985	1986	1987	1988	1989	1990	1991
Conservation headlands	52%	22%	60%	46%	39%	48%	25%	21%
Fully sprayed headlands	27%	13%	28%	22%	25%	30%	23%	18%
Number of farms	8	8	9	11	12	9	20	18

spraying of autumn crops against the aphid vectors of barley yellow dwarf virus). Agronomic costings have shown that financial losses incurred as a result of implementing Conservation Headlands are less than 1% (Boatman and Sotherton 1988).

Field experiments have demonstrated that the percentage weed cover in Conservation Headlands is over four times as high as in fully sprayed headlands, and that Conservation Headlands contain, on average, three times as many weed species (Sotherton 1991). In terms of insects, densities of the groups consumed by partridge chicks can be three times as high in Conservation Headlands than in fully sprayed headlands. The survival of partridge chicks follows suit: in each of eight experimental years, the survival rate was higher where Conservation Headlands were present than where they were absent (Table 8.1). With Conservation Headlands, it exceeded 30%, i.e. the minimum required to maintain a stable population (Potts 1986), in five of those years, whereas without them it reached that level in one year only.

8.4.2   Management of nesting habitat

With respect to nesting cover, traditional management has been the planting and maintenance of hedgerows on low grassy banks, thereby providing both cover and drainage at the same time (Anon. 1992). The hedges are kept below 2 m in height, and free of tall trees that could provide look-out perches for avian predators. The banks are cut every 2–3 years on a rotational basis, to avoid scrub encroachment while promoting nesting cover. Two new methods enable the control of pernicious weeds while safeguarding the vegetation that provides grey partridges with nesting sites. The first is the sterile strip, which is simply a 1-m buffer between the crop and the field margin that is kept free of weeds by rotavation or herbicide use. The buffer prevents invasion of the crop by weeds growing in the margin, and also insulates to some extent the margin from crop fertilizers and spray drift (Boatman and Wilson 1988). The second is the use of selective herbicides in the margin, which eliminates pernicious weeds without affecting the rest of the vegetation. Experimental trials have identified compounds such as fluazifop-P-butyl, which has a high specificity against brome, and quinmerac, which targets cleavers (Boatman 1992).

Where fields are large, additional nesting cover may be established by subdividing them in a non-permanent fashion using grass strips or 'Beetle Banks' (Rands 1987; Thomas *et al.* 1991). These are strips or slightly raised banks constructed across a field by repeated passes with a plough, and sown with tussock-forming grasses such as cocksfoot *Dactylis glomerata* or Yorkshire fog *Holcus lanatus*. The banks end 10 m away from the field margins to facilitate the use of farm machinery and prevent access to walkers. After a few years, the vegetation on Beetle Banks becomes too dense to be of much value; it is then easy to plough them up and build new ones.

8.4.3   Combined management of brood-rearing and nesting habitats

The techniques that cater for chick-food insects and nesting habitat should be combined for maximum benefit (Anon. 1992). Figure 8.8 illustrates schematically how, for instance, Conservation Headlands can be implemented next to suitable nesting cover in an agriculturally friendly way. In similar fashion, it is possible to have Conservation Headlands either side of a Beetle Bank. In either case, the juxtaposition of the two types of habitat means that nesting partridges can lead a newly hatched brood straight from the nest into brood-rearing areas.

The arrival of set-aside, which the European Union introduced in 1988 to

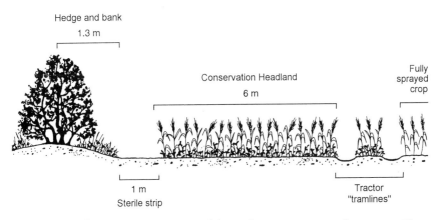

Figure 8.8. Schematic representation of latter-day management of grey partridge nesting habitat (hedge and grassy bank) combined with management of brood-rearing habitat (Conservation Headland).

reduce cereal surpluses, now offers many opportunities for partridge management (The Game Conservancy Trust 1993). Set-aside is the taking of land out of cereal production in return for continued cereal subsidies. In its initial voluntary five-year form, it was of little value because it rapidly became overgrown with dense perennial vegetation. It was replaced in 1992 with (effectively) mandatory rotational set-aside on 15% of cultivated area. This was potentially much more useful for game management because it could be established after the cereal harvest, by leaving the stubbles over winter and allowing vegetation to regenerate naturally. Overwinter stubbles could provide feeding grounds for grey partridges as well as leaving chick-food insects that overwinter in the soil undisturbed by cultivation. The regenerated vegetation could subsequently provide suitable brood-rearing habitat.

In practice, the set-aside regulations obliged farmers either to plough the land shortly after 1 May, or cut the vegetation at least once before 1 July, to control pernicious weeds. This proved disastrous for nesting birds (Poulsen and Sotherton 1992). Changes in the regulations in 1993 now enable the full wildlife potential of set-aside to be realized (The Game Conservancy Trust 1993). Cutting on rotational set-aside may be delayed until 15 July or replaced by herbicide treatment that preserves the vegetational structure. New 'flexible' set-aside, lasting a minimum of three years, allows the creation of gamebird habitats such as grass strips and banks for nesting, and the use of unharvestable crop mixtures such as cereals and kale for brood-rearing and the provision of food in winter. It is even possible to combine the two types of annual and longer-term set-aside to obtain an optimal configuration for game on each farm. The only remaining problem with the set-aside scheme is the uncertainty over its future. From an initial 15% of arable land, the percentage required to remain eligible for subsidies has been gradually reduced, and now stands at 5% for the 1996/97 growing season.

8.5 PREDATION MANAGEMENT

For centuries, the traditional role of the British gamekeeper has been to control predation on gamebirds and their eggs by killing predators. Trapping, shooting and poisoning predators was particularly widespread and intense during the 1800s, and resulted in the extinction of several raptor species and considerable reduction in numbers and range of many other avian and mammalian carnivores (Tapper 1992). Protective legislation during the twentieth century has restricted both the species that may be legally killed and the techniques that can be used to kill them (Anon. 1994). For instance, all raptors and owls have been protected since 1961 (and many of them well before that), while poisoning was outlawed already in 1911. The laws are not easy to enforce, and regrettably, the illegal killing of protected species still takes place albeit on a much reduced scale.

Plate 1 Crocodile eye reflections in camera flash. Most of the crocodiles in this farm pool have been captured as wild hatchlings and were located at night by villagers using torches. (Photograph courtesy of Mainland Holdings.)

Plate 2 *Crocodylus porosus* approaching culling size in a large commercial farm. (Photograph courtesy of Mainland Holdings.)

Plate 3 Peregrine falcons being reared for reintroduction to the wild in Germany. (Photograph courtesy of Christian Saar.)

Plate 4 An early (1977) experimental release point for German peregrines was the Tempelhof airfield in Berlin. (Photograph courtesy of Christian Saar.)

Plate 5 Mobile field abattoir used in the FAO Kenya Wildlife Management Project, 1971–1978.

Plate 6 Butchering wildebeest carcasses in the mobile field abattoir, Kenya.

Plate 7 Cape buffaloes in Mozambique have been killed primarily for their meat but these scrotums are being dried for use as tobacco pouches.

The question naturally arises whether nowadays predation control is still justifiable, not least because of the widely accepted ecological belief that predator numbers are controlled by the abundance of their prey and not vice versa – Errington's (1946, 1956) doomed 'surplus' theory. In fact, several scientific experiments have refuted that belief, and demonstrated convincingly that generalist predators can (and do) depress numbers of their gamebird prey. Thus the experimental removal of mammalian predators increased reproductive success, autumn abundance and spring density of tetraonids in Sweden (Marcström *et al.* 1988). Similarly in the case of the grey partridge, the experimental application of legal predation control in spring and early summer increased breeding success, post-breeding density, spring density and the shooting bag (Tapper *et al.* 1996); after three years, the autumn stock averaged 3.5 times higher when predation was controlled than when it was not, and the breeding stock 2.6 times higher.

The latter study is particularly important because it demonstrates that there is no need to break the law in order to achieve the benefits to game-birds of reducing predation pressure. The predators that may legally be killed are common, and several are increasing in abundance in the UK, namely foxes, small mustelids, rats, grey squirrels and corvids (Marchant *et al.* 1990; Corbet and Harris 1991). The highly seasonal predation control, carried out during the partridge nesting season, is directed at the period of greatest sensitivity to predation. The control methods that are currently legal are firearms (rifle or shotgun for foxes, rats, grey squirrels and corvids), lethal devices designed to kill instantly and whose location avoids the capture of non-target species (tunnel traps for rats and mustelids), and non-lethal trapping or restraining devices that allow non-target animals to be released unharmed (cage traps for corvids, mustelids and grey squirrels, stopped snares for foxes). Anticoagulant poisons are permitted for use against rats and grey squirrels, under strictly defined conditions.

8.6   THE PARTRIDGE ESTATE: AN INTEGRATED PACKAGE

Because the grey partridge is so prolific, it has the potential to respond quickly to a reduction in numbers incurred through shooting if its environment is suitable, in other words if predation is low, chick survival is high and nesting cover is plentiful. As discussed above, these are all features that can be manipulated by management. The population model developed by Potts (1986) can be used to investigate the relationships and interactions between stock size, harvest rate and management (Figure 8.9). In the currently prevailing unmanaged situation with modern intensive agriculture, the stock has little resilience to harvesting: a 20% harvest rate approximately halves spring density for little return, and a 50% one drives it to extinction. With full habitat management but no predation control, the equilibrium level for spring density in the absence of shooting increases

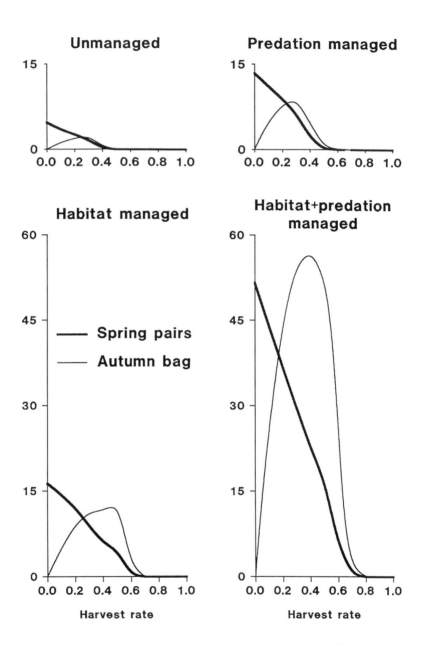

Figure 8.9 Stock/yield relationships for grey partridges in four different management situations: no management, i.e. the prevailing situation on intensive arable farms (top left), habitat management only (bottom left), predation management only (top right), and full management (bottom right).

from under 5 pairs/km² to 16 pairs/km². A 20% harvest rate decreases spring density by about a third, with a return of close to 10 birds/km². This situation is very similar to that with predation management but no habitat management. With full management of habitat and predation, the unharvested spring stock exceeds 50 pairs/km²; in this ideal partridge environment, the stock is highly resilient to shooting and, even at a 40% harvest rate, is still higher than the unharvested stock with habitat management alone. The bag is correspondingly high, in excess of 55 birds/km².

The economic implications of this modelling exercise are stark (Aebischer 1991b). Management of any kind costs money, and an estate will abandon its partridge shooting if it does not cover its costs. The truth is that habitat management alone does not provide a bag large enough to defray costs, so cannot be implemented in isolation. What about saving costs by ignoring habitat management and concentrating on predation control? Again, the bag is too low for cost-effectiveness. Only when habitat and predation management are combined is the result an integrated and self-supporting package with benefits to the partridge (more birds despite shooting than under any other scenario), to the estate owner (good shooting at no net cost) and to many other forms of non-game wildlife that, like the grey partridge, have suffered from the changes wrought by the intensification of agriculture (see next section). It was the unravelling of this package following a long-term reduction in chick survival rates and associated bags that led to the observed declines. Conversely, the remaining partridge estates with high stocks in the UK are ones where effective management of habitat and predation is considered an integral part of the approach to farming.

8.7 OTHER BENEFITS OF PARTRIDGE MANAGEMENT

Many other plant and animal species besides the grey partridge have been affected by agricultural intensification, ranging from the almost complete disappearance of certain arable weeds, to greater or lesser declines in number of invertebrates, small rodents, other game animals and farmland passerines (Smith 1986; Aebischer 1991a; Potts 1991; Fuller *et al.* 1995). Several of the management techniques outlined above have produced unexpected conservation benefits for such farmland wildlife, and also for the farmer in the form of biological pest control.

For example, rare annual arable weeds such as pheasant-eye *Adonis annua* and shepherd's needle *Scandix pecten-veneris* have appeared in conservation headlands, on sites where they had not been seen for many years (Sotherton 1991). The disappearance of such arable weeds was largely due to their susceptibility to herbicides, and many of them have survived only as seeds in the soil. The reduction in pesticide use but continued cultivation encouraged the seeds to germinate and flourish, to the extent that out of 25 rare species, 17 have been found in Conservation Headlands.

The proliferation of flowering plants in Conservation Headlands has in turn been found to attract butterflies and hoverflies, which seek nectar and pollen on which to feed (Sotherton 1991). Many of the butterflies use these energy sources for reproduction, and find their caterpillar host plants in the grassy banks and hedge bottoms managed as partridge nesting cover. Hoverflies do likewise, with the difference that they lay their eggs within the crop; the larvae that hatch from them are voracious predators of aphids and useful agents of biological control (Cowgill *et al.* 1993). So too are the predatory beetles that have been found to overwinter in vast numbers (up to 1500 beetles/m^2) in Beetle Banks, and disperse out into the surrounding crop in the spring (Thomas *et al.* 1991).

The greater weediness of Conservation Headlands leads to more weed seeds and to more insects. Small rodents have been found to forage preferentially in Conservation Headlands because of the increased seed resources (Tew *et al.* 1992). Many farmland passerines feed their chicks on insects, and they too can take advantage of the improved chick food availability. Research currently underway also suggests that several of the declining farmland passerines, like song thrush *Turdus philomelos* and linnet *Acanthis cannabina*, show improved nesting success when numbers of nest predators such as corvids and mustelids are reduced through predation management.

8.8   PROGNOSIS AND CONCLUSION

There is no indication in Figures 8.1 and 8.2 of a recent recovery in the fortunes of the grey partridge in the UK, despite the existence of tried and tested management methods. What, then, of the future? The problem is that partridge management costs money, time and effort, at least initially, so that nationally too few farmers are willing to deviate from the currently lucrative intensive farming methods. To achieve results on a national scale, agricultural subsidies need to be redirected from production towards environmental objectives in recognition of the ecological value of cropped land. The reform of the European Union's Common Agricultural Policy in 1992 provided a legislative framework that could be used to promote arable extensification, but its potential in the UK still remains largely untouched (Potts 1996). Its main benefit to date has been to allow wildlife-friendly forms of set-aside (section 8.4.3), but no financial incentive is offered to offset the extra work involved in implementing them, and set-aside may be phased out in the near future. What is needed is a scheme to foster arable extensification in the UK, as exists already in Spain – a plea backed by many British conservation organizations, so far to no avail.

At a local level, however, the conservation message is spreading. For example, in 1994 there were about 1920 km of Conservation Headlands and at least 98 km of Beetle Banks across the arable areas of the UK (Potts

1997). In Norfolk, a small group of concerned landowners and farmers started the Norfolk Partridge Group in 1987, first as a means of comparing notes, then, with The Game Conservancy Trust's involvement in 1991, of encouraging one another to manage land for partridges. Figure 8.10 shows that spring stocks tripled on estates which embraced the full partridge management package in 1992, compared with a fall of a third on nearby ones that did not. Within the last year, their success has encouraged the formation of other such county groups. It has been helped by the realization that, shooting opportunities now being extremely rare, the grey partridge commands a high premium than even red grouse *Lagopus lagopus scoticus*. Thus the economics of grey partridge management are improving, and the future prospects for the species are correspondingly good.

In conclusion, there are many lessons to be learnt from the principles underlying partridge management in the UK. Three main ones are summarized below.

The first is the apparently paradoxical result that shooting conserves grey

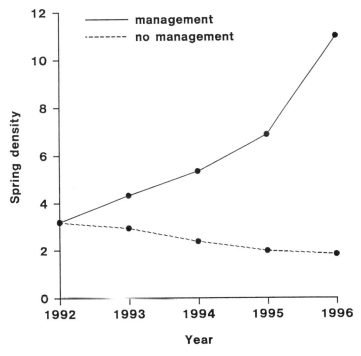

Figure 8.10 Average annual spring density (pairs/km^2) of grey partridges on five estates in Norfolk since full partridge management began in 1992, compared with that on five unmanaged estates from the same area.

partridges, by providing the incentive and the revenue to manage land appropriately. Throughout much of its range, this bird has declined drastically because of deterioration of its agricultural environment. Modelling shows that a cessation of shooting in a degraded environment will halt the decline only temporarily, but shooting and management together are able to reverse the decline and eventually produce a high sustainable density of partridges in a package that is effectively self-financing (Figure 8.11). This is the situation that is developing in Norfolk (see above, and Figure 8.10). For the vested shooting interest to produce this conservation success, it is necessary for it to have control of land management over a sufficiently wide area. This is the case in much of Britain, where estates are large, but is often not so in continental Europe, where cooperation between small farms is needed for effective management.

The second is the need for successful management to take into account

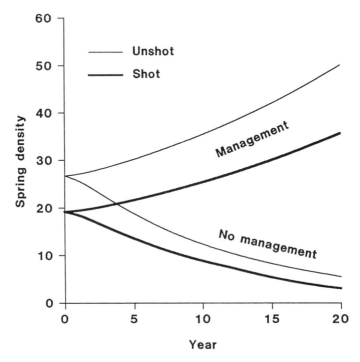

Figure 8.11 Simulated changes in abundance (pairs/km²) of grey partridges over 20 years, starting from a situation similar to Sussex in the early 1970s, in the case where lack of management leads to degradation, and in the one where efficient management leads to improvement, with the effect of shooting superimposed (from Aebischer 1991b).

modern agricultural requirements. The long-term survival of the grey partridge depends on the collaboration and support of farmers whose livelihoods are bound up with the land that they cultivate. Techniques such as Conservation Headlands and Beetle Banks are designed with these factors in mind, and aim to help the farmer through biological pest control and selective herbicide use without hindering his day-to-day activities.

The third is that the grey partridge is not the only farmland species to have suffered from agricultural intensification. As a result, the package of shooting and management that helps the grey partridge also helps many other forms of wildlife that live on cultivated land. Future agricultural policy should seek to encourage the extensive management of arable land that would promote farmland biodiversity generally.

REFERENCES

Aebischer, N.J. (1991a) Twenty years of monitoring invertebrates and weeds in cereal fields in Sussex, in *The Ecology of Temperate Cereal Fields* (eds L.G. Firbank, N. Carter, J.F. Darbyshire *et al.*). Blackwell Scientific Publications, Oxford, pp. 305–31.

Aebischer, N.J. (1991b) Sustainable yields: gamebirds as a harvestable resource. *Gibier Faune Sauvage*, 8, 335–51.

Aebischer, N.J. and Potts, G.R. (1994) Partridge *Perdix perdix*, in *Birds in Europe: Their Conservation Status* (eds G.M. Tucker and M.F. Heath). Birdlife International (Birdlife Conservation Series No. 3), Cambridge, pp. 220–1.

Anon. (1992) *Wild Partridge Management*. Game Conservancy Ltd, Fordingbridge.

Anon. (1994) *Predator Control*. Game Conservancy Ltd, Fordingbridge.

Batten, L.A., Bibby, C.J., Clement, P. *et al.* (1990) *Red Data Birds in Britain: Action for Rare, Threatened and Important Species*. T. & A.D. Poyser, London.

Boatman, N.D. (1992) Improvement of field margin habitat by selective control of annual weeds. *Aspects of Applied Biology*, 29, 431–6.

Boatman, N.D. and Sotherton, N.W. (1988) The agronomic consequences and costs of managing field margins for game and wildlife conservation. *Aspects of Applied Biology*, 17, 47–56.

Boatman, N.D. and Wilson, P.J. (1988) Field edge management for game and wildlife conservation. *Aspects of Applied Biology*, 16, 53–61.

Corbet, G.B. and Harris, S. (1991) *The Handbook of British Mammals*, 3rd edn. Blackwell Scientific Publications, Oxford.

Cowgill, S.E., Wratten, S.D. and Sotherton, N.W. (1993) The effect of weeds on the numbers of hoverfly (Diptera: Syrphidae) adults and the distribution and composition of their eggs in winter wheat. *Annals of Applied Biology*, 123, 499–515.

Errington, P.L. (1946) Predation and vertebrate populations. *Quarterly Review of Biology*, 21, 144–77, 221–45.

Errington, P.L. (1956) Factors limiting vertebrate populations. *Science*, 124, 304–7.

Ford, J., Chitty, H. and Middleton, A.D. (1938) The food of partridge chicks (*Perdix perdix* L.) in Great Britain. *Journal of Animal Ecology*, 7, 251–65.

Fuller, R.J., Gregory, R.D., Gibbons, D.W. *et al.* (1995) Population declines and range contractions among lowland farmland birds in Britain. *Conservation Biology*, 9, 1425–39.

Green, R.E. (1984) The feeding ecology and survival of partridge chicks (*Alectoris rufa* and *Perdix perdix*) on arable farmland in East Anglia. *Journal of Applied Ecology*, 21, 817–30.

Marchant, J.H., Hudson, R., Carter, S.P. *et al.* (1990) *Population Trends in British Breeding Birds*. British Trust for Ornithology, Tring.

Marcström, V., Kenward, R.E. and Engrén, E. (1988) The impact of predation on boreal tetraonids during vole cycles: an experimental study. *Journal of Animal Ecology*, 57, 859–72.

Potts, G.R. (1980) The effects of modern agriculture, nest predation and game management on the population ecology of partridges (*Perdix perdix* and *Alectoris rufa*). *Advances in Ecological Research*, 11, 1–79.

Potts, G.R. (1986) *The Partridge: Pesticides, Predation and Conservation*. Collins, London.

Potts, G.R. (1991) The environmental and ecological importance of cereal fields, in *The Ecology of Temperate Cereal Fields* (eds L.G. Firbank, N. Carter, J.F. Darbyshire *et al.*). Blackwell Scientific Publications, Oxford, pp. 3–21.

Potts, G.R. (1997) Cereal farming, pesticides and grey partridges, in *Farming and Birds in Europe* (eds D.J. Pain and M.W. Pienkowski). Academic Press, London, pp. 150–77.

Potts, G.R. and Aebischer, N.J. (1991) Modelling the population dynamics of the grey partridge: conservation and management, in *Bird Population Studies: Their Relevance to Conservation Management* (eds C.M. Perrins, J.-D. Lebreton and G.J.M. Hirons). Oxford University Press, Oxford, pp. 373–90.

Potts, G.R. and Aebischer, N.J. (1995) Population dynamics of the grey partridge *Perdix perdix* 1793–1993: monitoring, modelling and management. *Ibis*, 137, Supplement 1, 29–37.

Poulsen, J.G. and Sotherton, N.W. (1992) Crow predation in recently cut set-aside land. *British Birds*, 85, 674–5.

Rands, M.R.W. (1987) Hedgerow management for the conservation of partridges *Perdix perdix* and *Alectoris rufa*. *Biological Conservation*, 40, 127–39.

Smith, A. (1986) *Endangered Species of Disturbed Habitats*. Nature Conservancy Council, Peterborough.

Sotherton, N.W. (1991) Conservation Headlands: a practical combination of intensive cereal farming and conservation, in *The Ecology of Temperate Cereal Fields* (eds L.G. Firbank, N. Carter, J.F. Derbyshire *et al.*). Blackwell Scientific Publications, Oxford, pp. 373–97.

Sotherton, N.W., Robertson, P.A. and Dowell, S.D. (1993) Manipulating pesticide use to increase the production of wild game birds in Britain, in *Quail III: National Quail Symposium* (eds K.E. Church and T.V. Dailey). Kansas Department of Wildlife and Parks, Pratt, pp. 92–101.

Southwood, T.R.E. and Cross, D.J. (1969) The ecology of the partridge III. Breeding success and the abundance of insects in natural habitats. *Journal of Animal Ecology*, 38, 497–509.

Tapper, S.C. (1992) *Game Heritage: An Ecological Review from Shooting and Gamekeeping Records*. Game Conservancy Limited, Fordingbridge.

Tapper, S.C., Potts, G.R. and Brockless, M.H. (1996) The effect of an experimental reduction in predation pressure on the breeding success and population density of grey partridges (*Perdix perdix*). *Journal of Applied Ecology*, **33**, 965–78.

Tew, T.E. MacDonald, D.W. and Rands, M.R.W. (1992) Herbicide application affects microhabitat use by arable wood mice (*Apodemus sylvaticus*). *Journal of Applied Ecology*, **29**, 532–9.

The Game Conservancy Trust (1993) *Factsheet 3. Game, Set-aside and Match.* The Game Conservancy Trust, Fordingbridge.

Thomas, M.B., Wratten, S.D. and Sotherton, N.W. (1991) Creation of 'island' habitats in farmland to manipulate populations of beneficial arthropods: predator densities and emigration. *Journal of Applied Ecology*, **28**, 906–17.

9

Birds of prey and modern falconry

9.1 INTRODUCTION

Falconry, or hawking, is the taking of quarry in its wild state by using trained birds of prey. Generally regarded as a sport, it seems probable that it began as a way of filling the cooking-pot in the way that a few people in China still use trained cormorants to catch fish. The origins of falconry are obscure but the earliest records are all from parts of Asia and indicate that birds of prey were being trained at least 2500 years ago. Falconry had reached Europe (and, presumably, the Arab countries) by the fifth or sixth century (De Bastyai 1973) and since then, even in Europe, it has never completely died out.

Hawking had a very strong following in medieval Europe but the invention of the 'fowling piece', as the seventeenth-century shotguns were called, provided an alternative sport with more popular appeal. Largely as a result of this, European falconry became the pursuit of small minorities, banded together in clubs and associations to keep their sport alive.

9.2 THE BASICS OF FALCONRY

There are several recent books on the subject (e.g. Woodford 1971; Parry-Jones 1988; Durman-Walters 1994) and there is neither space nor need for more than an outline here.

9.2.1 Taming and training

Food is the key and the falconer must convince the hawk that a falconer's glove is a safe place to feed. Eventually, it will fly to the fist when called. Gradually, the hawk is made familiar with strangers, vehicles and all the sights and sounds that a trained bird is likely to encounter. When necessary,

Conservation and the Use of Wildlife Resources. Edited by M. Bolton.
Published in 1997 by Chapman & Hall. ISBN 0 412 71350 0.

a hood can be used, like the blinkers on a horse, to prevent a hawk from being upset or distracted.

The falconer needs particular skill and patience with hawks that have been trapped as mature adults (haggards). Hawks taken from the nest (eyasses) may be somewhat easier but because they have never had to fend for themselves they will take longer, when trained, to develop their hunting skills and to show their individual character and abilities.

Falconers have found that individual hawks (within a species) can differ in temperament and behaviour as much as do dogs. In traditional Arab falconry the falcons were trapped on migration and released again after the hunting season. Only outstanding birds would be kept through the moult to be used again the following season.

Hawks can become tame to a fault if they are hand-reared from very young nestlings because imprinting can result in their continuing to regard humans as parents and then as potential mates. They can be noisy (screamers) and aggressive. Imprinted birds cannot be expected to perform normal courtship and mating with their own kind.

In order to permit fledgling hawks to gain flying fitness and experience before they are taken up for training, they may be allowed to fly free at hack. Hacking is only feasible in places where the birds are not likely to be harmed or seriously disturbed by people. Food is left for them every day on a hack board at the point of release. As days go by, the young hawks range over greater distances but, because they have no natural parents to feed them, they keep returning to the hack board until the time when they are fully independent. Before that point is reached, the falconer will terminate hack by recapturing the hawks, usually with a net at the hack board.

9.2.2   Hawks and hawking

It is convenient and customary to refer to all falconers' birds as hawks and the practice of hunting with them is called hawking. But when falconers need to be a little more specific, they will use the word 'longwing' to refer to the true falcons (genus *Falco*) whereas hawks, typified by the genus *Accipiter*, are collectively called shortwings. In general, longwings are birds of open country and they hunt by stooping on their prey from a height, whereas shortwings can be flown in enclosed or even wooded country and flights tend to be closer to the ground and more of a chase.

Eagles, though not commonly trained, are flown from the fist as are birds of the genus *Buteo* and *Parabuteo* – collectively referred to as broadwings. These birds tend to be slower than true hawks and falcons but are used successfully in enclosed country. Harris hawks (*Parabuteo unicinctus*) and red-tailed hawks (*Buteo jamaicensis*) are now popular with falconers on both sides of the Atlantic.

Trained hawks do not retrieve and the falconer must get to a hawk and take it off a kill before it can satisfy its appetite. Otherwise there will be no more flights that day and the hawk may fly off and settle where it chooses to digest its meal. The falconer is then at risk of losing it completely if it is not recovered quickly in the days that follow. It is customary to use a lure to bring a longwing back to the falconer. The hawk soon learns to stoop on any object, garnished with meat and swung on a cord.

For centuries, bells have been used as a means of locating hawks in cover but modern technology has made available miniature radio transmitters which can be fitted to the legs or tail of even the smallest hawks. Birds can also be implanted with passive transponders (identichips) for positive identification. Nevertheless, hawks are still occasionally lost and it is part of the falconers' code of good conduct to fit hawks with special anklets and slitless leg-straps (jesses) when they are being flown. There is far less risk of a lost hawk becoming caught up in vegetation or on barbed wire if the jesses are free to slide out of the anklets and there are no open slits. Traditional jesses, with slits for swivel attachment, need be used only to tether a hawk at a perch or block. A tethered hawk, in contrast to a bird in a cage, cannot easily damage its feathers. For falconry, perfect feathers are important, as is the hawk's flying weight and condition. It is good practice to weigh hawks regularly and maintain their flying weight during the hawking season, but they should be fed to repletion and kept untethered in a special enclosure (mews) during the moult.

9.3 THE FIRST HALF OF THE TWENTIETH CENTURY

By the turn of the century, in western countries at least, falconry was a sport for the very few. In Great Britain the Old Hawking Club, founded with seven members in 1864, had a membership of 16 by the year 1900. In Germany the Deutscher Falkenorden was founded in 1923 and produced the world's first magazine on falconry. The British Falconers' Club was founded four years later, with 15 members, following the winding up of the Old Hawking Club in 1926.

Elsewhere, in Europe, the colonies and the USA, falconry was pursued by individuals and small groups under circumstances which, except for the war years, were in some ways easier for the sport than they are today. Hawks, still widely classed as vermin, could be taken from the wild and flown at quarry that was plentiful and in countryside that was relatively quiet.

In 1954 the British Protection of Birds Act was welcomed by the British Falconers' Club (membership then stood at 90) as it protected nearly all raptors but made provision for falconry to continue under licence. But shortly after that, for reasons not then understood, raptor populations began to decline dramatically. In 1958, an article on the lack of breeding success in wild peregrines in the west of England appeared in the British Falconers' Club journal (Treleaven 1958). Similar reports from other parts of the United Kingdom confirmed that very few peregrines were nesting successfully.

9.4   THE PESTICIDE CRISIS

The publication of *Silent Spring* (Carson 1962) perhaps did more than anything else to arouse public concern. In the UK a voluntary ban was accepted on most uses of aldrin, dieldrin and heptachlor, and research was begun in 1964 at Monks Wood Experimental Station into the effects of organochlorine pesticides on birds of prey. But the damage was done; in the UK, nesting peregrines were reduced to a third of their pre-war numbers. In the USA they were exterminated as a nesting species east of the Mississippi (Cade 1985) and became the first animal in the USA to be classed as endangered. In the former West Germany the breeding population fell from at least 350 pairs to about 50 (Speer 1985). In East Germany no breeding peregrines survived. A comparable situation existed in many other countries. The pesticide story is now well known and the use of persistent poisons, which accumulate in food chains, has at last been banned or strictly controlled in many countries, although in some parts of the world there remains cause for serious concern. India, for example, still manufactures and uses thousands of tonnes of DDT (Gupta 1986).

Possibly as a result of publicity about raptors, the popularity of falconry and the keeping of hawks by non-falconers began to increase in Europe and the USA. The timing was particularly unfortunate. By the end of the 1960s the demand for birds of prey could not be met and a number of people had been prosecuted for illegally taking young birds and eggs from the wild.

9.5   RESPONSES AND RECOVERIES

As well as sharing the general concern about the mass poisoning of wildlife, falconers were confronted with three major problems affecting their sport:

- a grave shortage of some species of wild hawks
- a damaged reputation resulting from bad publicity caused by law-breakers
- increasing demands for more legal restrictions on falconry and the taking of hawks

Their responses will now be discussed.

9.5.1   Captive breeding

One might suppose that falcons, in the past, had occasionally bred in captivity but captive breeding was not known to living falconers until 1942. In that year a German falconer, the late Renz Waller, bred a peregrine in his garden near Dusseldorf. During the 1960s the need to breed hawks in captivity became urgent and on both sides of the Atlantic falconers began trying to convert their mews into what they hoped would become breeding enclosures instead of mere moulting sheds.

Successes came slowly at first but in 1970 a group of falconers led by Tom Cade of Cornell University established a major facility for peregrine breeding which, through the Peregrine Fund, was to become the renowned World Center for Birds of Prey. Throughout the USA and Europe, falconers and other enthusiasts also pooled information on a variety of species. The first captive breeding of sparrowhawks (*Accipiter nisus*) in the UK, for example, occurred in a falconer's suburban back garden, and details of the enclosure were widely adopted by other breeders (Hurrell 1973).

During the 1970s the basic requirements became well known and captive-bred hawks were soon being flown in Europe and America. There is no one correct procedure but the following factors proved to be key elements in breeding success:

- careful matching of potential mates with regard to disposition and prior experience
- seclusion and prevention of disturbance to the breeding enclosures of mated pairs
- regulation of physical condition by diet and general care
- special attention to imprinted birds

The problem of imprinted birds is complex. Breeding behaviour largely depends on whether imprinting to humans has occurred in the male, the female or both birds of a pair. Under some circumstance imprinting can be reversed but where birds strongly prefer humans as potential mates it can be more productive to encourage the preference in order to achieve voluntary artificial insemination (Nelson 1977).

AI has been carried out in poultry since the 1930s but was not developed as a technique in raptors until the 1970s. Collecting semen from a male bird needs at least two people: one must hold the bird while another massages it to express the semen, which appears at the cloaca. The semen is transferred by syringe into the oviduct of the female. The female must also be held and the oviduct everted for this operation. The procedure is described by Weaver (1985a).

The advantage of using imprinted hawks for artificial insemination is that it is possible to get males to copulate with objects such as a hat, modified for semen collection, and females will present the cloaca for insemination by a syringe. This method was pioneered by falconers in the early 1970s (Hamerstrom 1970; Berry 1972; Grier *et al.* 1972; Boyd 1978). Fertile eggs obtained by this method can be hatched in an incubator or under foster parents – including bantam hens.

On the use of imprinted female peregrines for this purpose, however, Weaver (1985b) remarked that it may be too labour intensive to be of use in a large operation, but that it is 'eminently suited to the individual fal-coners from whom the technique is borrowed. Falconry offers the means to keep the bird in the best of physical and mental condition'. Similarly, Boyd and Schwartz (1985) point out that artificial pair-bonds between humans and male peregrines are most easily maintained by using the 'traditional routines of falconry'.

More than 30 species of diurnal raptors, world wide, have now been bred for at least two generations (F_2) in captivity. Bloodline peregrines are now available, complete with pedigree forms.

9.5.2　Political representation

Conscious of the various socio-political threats to their sport, falconers in the 1960s gradually became better organized. In America, before 1964, there had been no single representative body but in that year the North American Falconers' Association was formed. Four years later, in Germany, the International Association for Falconry and Conservation of Birds of Prey (IAF) was created and the constituent assembly elected a member of the British Falconers' Club as its first president. Falconry could now be represented by a participating delegate at meetings of such bodies as the International Council for Bird Preservation (ICBP), IUCN, and in due course, CITES. The IAF now represents more than 20 member states.

9.5.3　Improvement of public image

In order to acquire and maintain a good image, concerned falconers, through their clubs and associations, were at pains to introduce codes of conduct and raptor welfare standards and to ensure that newcomers to the sport were made aware of the time and commitment that they would need. But stringent controls could not always be enforced. Birds of prey were being popularized by the media, were increasingly on display at country shows and tourist exhibitions and were becoming available from commer-cial breeders. In some countries compulsory membership of a falconry asso-ciation, involving a progression through a sort of falconer's apprenticeship, was built into the legal framework of the sport, but in other countries (including the UK) membership of any association is entirely voluntary.

The link between conservation and falconry was perhaps most easily established in America and southern Africa – where falconers were heavily involved in publicized activities such as bird-banding and peregrine hacking.

9.6　THE PRESENT SITUATION

In the west some wild raptor populations have more than recovered from the pesticide crisis. Peregrines in the UK now have a breeding population of about 1200 pairs compared with the 850 pairs believed to have existed pre-war. They are regarded as a menace by those who race pigeons. Wild kestrels and sparrowhawks are at capacity level and merlins are occupying about 90% of their former range. Seven other species are still well below capacity, with deliberate persecution reported to be the main limiting factor (Newton 1994). Peregrine populations in the USA have recovered to the extent that permits to take birds from the wild are again being issued.

There must now be more people involved with birds of prey than at any time since the Middle Ages. In Britain alone, in 1994, there were some 16 000 birds of prey (Table 9.1) registered, to about 8000 keepers, though probably less than a quarter of them regularly fly hawks at quarry. In continental Europe (excluding the former Soviet Georgia, where it is traditional to trap sparrowhawks but release them after a month's flying at migrating quail) there are an estimated 2500–3000 active falconers (C. de Coune, pers. comm.). In the USA the North American Falconers' Association has about 3000 members, which probably include the majority of practising falconers.

9.6.1 Falconry in the USA

Nearly all large longwings currently in use are captive-bred but American falconers obtain most other hawks from the wild (J. Hegan, pers. comm.). Legislation was introduced federally in 1978 which, among other things:

- established three progressive classes of falconry permit: apprentice, general and master
- required applicants to submit to a written examination on raptors and falconry

Table 9.1 Birds of prey registered with the Department of Environment, UK, in 1993 (source: Cromie and Nicholls 1995)

Species	Number	Species	Number
Kestrel *Falco tinnunculus*	3855	Sparrowhawk *Accipiter nisus*	1157
Common buzzard *Buteo buteo*	2472	Red-tailed hawk *Buteo jamaicensis*	1062
Harris hawk *Parabuteo unicinctus*	1815	Goshawk *Accipiter gentilis*	880
Peregrine *Falco peregrinus*	1418	Merlin *Falco columbarius*	361
Other species (approximate number)			3000

- named the species and numbers of raptors which could be taken from the wild and possessed
- set standards for general care of hawks

In 1983, because of the growth of interest in captive breeding, a separate set of regulations were enacted to cover raptor propagation. These regulations established standards but also made provision for the sale, purchase and barter of captive-bred birds of prey.

In October 1989, simplified and slightly more relaxed rules came into effect. Separate permits are required for falconry and raptor propagation but falconers, as well as breeders, are allowed to buy and sell captive-bred hawks that are marked with a numbered band supplied by the US Fish and Wildlife Services (USFWS). With the exception of two or three states, where there is no apparent interest, falconry is now regulated throughout the USA in accordance with the Federal Regulations. In promulgating the 1989 regulations the USFWS went on record as:

> Supporting falconry as a legitimate and lawful use of the migratory bird resource to the extent it does not adversely affect that resource . . . It has been the Service's experience that the overwhelming majority of falconers practice their sport in full compliance with Federal and State regulations. The Service also recognizes that most falconers are conservationists who have a deep and abiding love for the migratory bird resource. Further, the Service feels that falconry, raptor propagation, and migratory bird rehabilitation often serve as vital tools for public education about the need for conservation of raptors and other migratory birds in North America. (*Federal Register*, 54 (177), 14 September 1989.)

9.6.2   Falconry in the United Kingdom

In Britain the 1954 legislation was replace by the Wildlife and Countryside Act of 1981. Under this Act all birds of prey continue to be protected, and licences can be granted for falconry and aviculture. Falcons and hawks had to be ringed and registered with the Department of Environment (DoE) if kept in captivity but registration for common species was discontinued in 1994. Birds of prey bred in captivity, if ringed, may also be bought and sold under licence. A voluntary panel, the Hawk Board, represents all hawk-keeping interests in the UK and maintains liaison with the DoE on matters of licensing, inspections, etc.

The Act does not make provision for separate classes of falconry licence, nor does it require applicants to be examined or their facilities inspected before a licence is issued. The major falconry and captive breeding associations have produced codes of conduct for members, and the British Falconers' Club operates a scheme of graduated membership by which established members assist newcomers to gain experience.

No separate permit is needed to allow captive birds to be bred. Additional licences may be issued to take, with trained merlins and sparrowhawks, certain traditional quarry, such as blackbirds, which are common but protected (section 9.7.2). No additional licence is needed to take game and unprotected species such as crows and rabbits.

9.6.3 Falconry and the law in continental Europe

The provisions which have been made for falconry in Europe are much too varied to be describe here. They include quite different rules with regard to numbers and species of hawks which may be kept; which raptors (if any) may be taken from the wild; whether there is a competence test and whether this differs from a standard hunter's test; whether a falconer must be a member of a recognized falconers' association, and (in one country) the separation of falconry and shooting according to days of the week (C. de Coune, pers. comm.)

9.7 CONSERVATION CONSIDERATIONS

The most direct conservation aspects of falconry are now discussed.

9.7.1 Harvest of hawks from the wild

Small numbers of birds of prey, well below sustainable yields, are being taken from the wild for falconry in western countries. Permits for eyasses are issued more readily than for older birds as this is the most expendable segment of raptor populations. In all cases natural mortality is likely to be highest among young, non-breeding birds.

Radio telemetry has revealed possible upward biases in wild bird mortality estimates which are based on ring recovery. Radio-tagging can be used together with ringing studies to correct such biases. From radio-tagged goshawks in Sweden, Kenward (1993) estimated the first-year mortality rate of males and females to be 51% and 34% respectively. The estimate for adults for both sexes was 21%. A shorter, as yet unpublished, study on radio-tagged buzzards in England suggests a first-year mortality of about 33% (R. Kenward, pers. comm.).

Of the goshawks which died naturally, and for which proper autopsy was possible, 37% had died from starvation, suggesting that natural mortality in goshawks could be density-dependent – as it is for sparrowhawks (Kenward *et al.* 1991; Wyllie and Newton 1991). Significant numbers of raptors are still shot and poisoned indiscriminately.

Harvests of wild hawks for falconry are likely to be only a tiny fraction of sustainable yields. An offtake of 20–30% has been estimated as the sustainable yield for healthy peregrine and goshawk populations (Kenward

1987). In the USA a legal harvest of less than 1000 raptors a year, plus a smaller estimated illegal offtake, was considered to have 'no impact on the raptor resource' (USFWS 1988). Captive breeding has reduced the demand for wild hawks, although continued input of wild blood to some captive stocks is desirable for genetic reasons (Chapter 13).

9.7.2   Harvest of quarry species from the wild

Obviously, falconry does not compare with shooting as a method of filling the game-bag. The average falconer kills about 30 head of quarry a year (Fox 1995). Trained hawks are allowed to kill less frequently than they would need to do if they were free. Common protected birds, killed in Britain by trained hawks, are insignificant in number. For the years 1990–1993, for example, the mean annual kill was 548 birds of four species from an estimated population of about 10 million (Cromie and Nicholls 1995). Even allowing for illegal and unreported kills the toll of wild birds from falconry is negligible.

9.7.3   Losses of falconers' hawks

Table 9.2 shows some of the birds declared lost to the DoE in Britain during the 15-month period January 1991 to March 1992. In May 1994 the more common species – including the kestrel, common buzzard, sparrow-hawk, lanner and lugger – were removed from Schedule 4 of the Wildlife and Countryside Act 1981. As these species are no longer required to be registered, losses no longer need to be reported to the DoE. Birds of other species, reported lost when registration was due for renewal during 1993–4, are included in the table.

In general, assuming a hawk is lost in suitable country for the species (which ought to be the case for native birds being flown at quarry), survival could be expected to depend largely on the bird's condition at the time of loss, previous hunting experience and on the time it takes to lose its tameness in the presence of humans.

Kenward (1981) analysed the fate of goshawks imported from Finland for British falconers between 1970 and 1978. Of 33 lost eyasses 13 (39%) were known to have survived for at least 20 days. Two were found dead within the 20-day period and the remaining 18 were not heard of again. In an experimental release of radio-tagged goshawks four eyasses out of 10 survived (and were monitored for 18–35 days) during winter in unfamiliar country.

Using 40% as a conservative estimate for survival of eyasses after loss, then at least 13% of imported eyass goshawks would have successfully re-entered the wild in Britain – an average of 15 birds a year between 1970 and 1975 (Kenward 1981). In addition, goshawks were released deliberately

Table 9.2 Numbers of birds of prey (not all species) declared lost to the Department of the Environment, UK, January 1991–March 1992 (source: DoE statistics published in *Birding World*, March 1993)

Species	1991/92	1993/94*	Hybrid	1991/92	1993/94
Peregrine F. peregrinus	43	45	Peregrine × prairie falcon	3	–
Lanner F. biarmicus	32	–	Peregrine × saker	3	3
Goshawk A. gentilis	31	40	Saker × lanner/lugger†	2	–
Saker F. cherrug	19	–	Lanner × lugger	1	–
Prairie falcon F. mexicanus	4	–	Peregrine × gyr	1	1
Gyr falcon F. rusticolus	1	1	Peregrine × lanner/lugger	–	1
Hobby F. subbuteo	–	2	Gyr × saker	–	1
Merlin F. columbarius	–	16	Gyr/saker × saker	–	2

*Birds declared lost in 1993–1994 include only types which were still required to be registered following a relaxation of registration requirements in 1994 (source: DoE).
†F. jugger.

(section 9.7.5). Survival of longwings is not likely to be inferior to that of goshawks and there have been numerous sightings of ex-falconers' birds living wild.

9.7.4   The problem of exotics and hybrids

It remains to be seen whether any exotic raptor will become established in the wild from lost falconers' birds. There is no evidence, so far, of this happening – even though lanners (*Falco biarmicus*) have been used by British falconers since medieval times. Robert Kenward has suggested that in the UK only red-tailed hawks may present any risk (reported in Cromie and Nicholls 1995).

When captive breeding became common practice it was inevitable that some hawk breeders would create hybrids. To begin with the motive may have been nothing more than curiosity but falconers were soon seeking to improve performance in 'working birds' just as breeders have always done with dogs and horses. It is extremely unlikely that any hybrid could be more successful in the wild than are the existing products of natural selection. But hybrids are not expected to compete, long-term, against the forms selected by nature in a particular niche; they are simply required to hunt for a while under the circumstances that the falconer provides. For example, in country that is not open enough for peregrines it may be possible to get better quality flights from a peregrine × saker hybrid than from a bird of either species. Hybrids are particularly popular among American falconers.

It is to be expected that sympatric species will have evolved reproductive isolation mechanisms but types which do not naturally come into contact (allopatric species) will not have evolved the mechanisms which keep them apart. In captivity some closely related falcons, such as gyrs (*F. rusticolus*) and sakers (*F. cherrug*), will hybridize and are fully fertile over several generations. Other hybrids must be produced by AI and the more outcrossed combinations, such as Harris × redtail (*Parabuteo unicinctus* × *Buteo jamaicensis*), are probably infertile (Fox 1995). Nevertheless, Cromie and Nicholls (1995) follow the Raptor Research Foundation in recommending that hybrids between one or more allopatric species should not be bred for falconry, and other hybrids flown free should be imprinted on humans to reduce the risk of interbreeding with wild birds.

9.7.5 Application of falconry to deliberate reintroductions

Release techniques for birds of prey are probably better developed than for any other group (Wilson and Stanley Price 1994). Hacking and other falconers' practices can be used equally well for producing strong, skilled birds for falconry or for preparing inexperienced birds for life in the wild. When captive-bred birds are being reared for release to the wild, keepers do well to keep out of sight as far as practicable to avoid imprinting young birds to humans. Glove puppets, in imitation of the parent, have been used for hand feeding nestlings through a screen.

Fledged birds can be hacked to the wild as described in section 9.2.1. Alternatively, young can sometimes be placed with foster parents which happen to be rearing offspring of the same age. With foster parents of a different species (cross-fostering) there might be a problem of cross-species imprinting which could later interfere with mate selection (Bird *et al.* 1985). Possible dietary problems have also been reported with cross-fostering (Sherrod *et al.* 1982).

All three release techniques, hacking, fostering and cross-fostering, have been used in different parts of the world for a number of species in projects large and small. Some notable reintroduction programmes, in which falconers and falconry techniques have been involved, are very briefly reviewed below.

Table 9.3 Peregrine falcon releases and population recoveries in the USA (source: Peregrine Fund Annual Report, 1993)

Region	Known pairs	Individuals released*
Western States	>700	2598
Midwest and South Central Canada	57	698
Eastern States	98	1229

*Totals include releases by the Santa Cruz Predatory Bird Research Group, the Midwestern Peregrine Recovery Group, and the Canadian Wildlife Service.

(a) Peregrines in the USA and Canada

Between 1976 and 1982 the Peregrine Fund released nearly 2000 peregrines in 27 states of the USA (Sherrod *et al.* 1982). By 1993 the species was maintaining a wild population of about 100 breeding pairs in the eastern states where there were none in the 1970s. The status of peregrines in North America in 1993 is summarized in Table 9.3.

(b) Peregrines in Europe

The British Isles, a stronghold of *Falco peregrinus*, supported a viable population for recovery throughout the 'pesticide era' and few deliberate releases were made. In other parts of Europe the situation was more critical. In the former West Germany a release programme was initiated at Hessen in 1978 and, in a remarkable effort by a very few individuals funded by the Deutscher Falkenorden, a hundred peregrines had been released four years later (Saar 1985). To date, about 700 birds have been released and from them at least 70 nesting pairs have become established (Saar, pers. comm.) (Plates 3 and 4).

In Sweden captive breeding got off to a slow start in 1974 with the aim of building up a breeding stock in captivity. The first reintroductions were hacked back to the wild in 1982 (Lindberg 1985). A small breeding and reintroduction programme was started in France with the first two young being fostered to wild parents in 1975 (Monneret 1977).

(c) Goshawks in the United Kingdom

The European goshawk (*Accipiter gentilis*) was extinct in Britain by the end of the last century. It was reintroduced deliberately in the 1960s by falconers who discreetly established small wild populations in the north of England and near the Welsh border. The birds were imported from central

Europe until 1970 when import restrictions terminated the project. More goshawks were acquired for release, however, during the 1970s. These were obtained under licence and imported from Finland where they would otherwise have been killed by gamekeepers. They were used, together with some trained hawks, to establish a breeding nucleus in the English midlands (Kenward 1977, 1993).

The goshawk is now re-established as a British bird, with an estimated population of about 300 pairs in four main population centres.

(d) Sea eagles in the United Kingdom

By the beginning of this century the white-tailed sea eagle (*Haliaeëtus albicilla*) had been exterminated in Scotland and Ireland where it formerly occurred on the north and west coasts. Under special permits nestlings were obtained from Norway and hacked back to the wild from the Island of Rum Nature Reserve in the Scottish Hebrides. The work was initiated by the Nature Conservancy Council (NCC) and the Royal Society for the Protection of Birds (RSPB). A total of 82 birds were released between 1975 and 1985. The young eagles were fitted with jesses and a leash and tethered in an arrangement which permitted flight between a wooden shelter and a perching block. The arrangement was described in a German publication in 1974 (Trommer 1992). Immediately before release each eagle was fitted with a hood to keep it calm while measurements and marking were carried out (Evans *et al.* 1994).

In 1992 about 20 individuals were known to be occupying territories and seven young were fledged. Illegal poison baits (for foxes and crows) are a major threat to the eagles and the population is still too small and localized to be considered secure.

(e) Red kites in the United Kingdom

The red kite (*Milvus milvus*), once a common scavenger around English towns, survived into the twentieth century as a remnant population of only a few pairs in central Wales. Nest protection (mainly against egg collectors) enabled the species to survive and slowly increase its numbers to 78 pairs by 1992. However, the Welsh population remains isolated and vulnerable, with little opportunity for expanding its range (Lovegrove 1990).

A reintroduction programme was started in 1989 with 10 young birds from Sweden which were hacked to the wild at Scottish and English centres. The project was coordinated by the Joint Nature Conservation Committee (JNCC) and the RSPB. By 1994 a total of 186 red kites had been released; the numbers divided equally between the Scottish and English sites. Most of the birds were obtained as nestlings from Sweden and Spain and reared in groups of up to four until the age of about 12 weeks when they were released from aviaries at hack sites. Human contact was kept to a minimum. Some eggs were taken from deserted or otherwise doomed Welsh nests and these were hatched and reared under the supervision of a local expert falconer.

As in the case of the eagles, poisoning is a serious threat. Figures from the JNCC showed a total of 25 red kites found poisoned between 1979 and 1990, along with 255 buzzards and 28 golden eagles for the same period. But more than 70% of the birds released in England and Scotland were still alive in 1994 and were breeding in both regions for the first time in more than a century (Evans *et al.* 1994).

(f) Mauritius kestrel

Once the rarest raptor in the world, the Mauritius kestrel (*Falco punctatus*) was believed to have a wild breeding population of only two pairs in 1974. The combination of habitat destruction and pesticides had proved too much for this specialized little bird (Jones and Owadally 1981).

Captive breeding and rehabilitation was begun in 1973 with support from the ICBP and WWF. The species was bred in captivity and, in addition, eggs were harvested from the wild and hatched in incubators, or under foster parents, and hand-reared by a visiting specialist from the Peregrine Fund. Hatching and rearing rates were superior to those in the wild. The wild parents, whose eggs were harvested, would produce a second clutch.

At about 10 days old, young kestrels could be returned to wild foster parents when these were available but most were hacked to the wild from artificial nests in groups of two to seven at the age of about 1 month. A small captive breeding colony has also been established at the World Centre for Birds of Prey, and young are sent back to Mauritius courtesy of British Airways (Jones 1990).

By 1993 a total of 331 young had been returned to the wild and over 60 wild breeding pairs had become established (Anon. 1993). The field project was directed by a biologist/falconer and supported by the Government of Mauritius, the Mauritius Wildlife Appeal Found and the Jersey Wildlife Preservation Trust.

9.8 CONCLUSION

In their commissioned report to the RSPCA (UK), Cromie and Nicholls (1995) recognize the 'vital' conservation role of the people involved in coordinated captive breeding of raptors. They emphasize, however, that raptor keeping is no longer confined to falconry or aviculture but has become a widespread business involving a diversity of activities at public exhibitions, raptor centres, and on courses sometimes run by 'instant experts' whose standards of hawk management are less than satisfactory. Cromie and Nicholls call for a body such as the Hawk Board to act as quality auditors in approving the status of such centres and courses. Approved clubs, it is suggested, should run apprenticeship schemes such as that introduced by the

British Falconers' Club, and consideration should be given to making approved club membership compulsory in the UK.

Many falconers hope that the present popularity of hawk keeping will not be sustained for it presents a dilemma. There is no benefit to falconry, or to hawks, in the persistence of romantic images if the imagery attracts people to the sport when they have neither time nor facilities to do justice to it. Nor is there any value to falconry, conservation, or aviculture, in raptors becoming popular pets. On the other hand, those in falconry who try to discourage newcomers are at risk of driving novices into the hands of those with lower standards.

Recent trends notwithstanding, serious falconers and raptor breeders now represent a combination of enthusiasm and expertise that has no precedent. Its value to the conservation of wild raptors has been amply demonstrated and is generally recognized. Raptors have never ceased to be at risk from shooting, poisoning, nest destruction and habitat loss, and conservation will be less well served if falconry comes to be entirely associated with the production of domestic hunting birds. It is through its association with wild raptor populations that modern falconry makes it greatest contribution to conservation.

ACKNOWLEDGEMENTS

I owe special thanks to Nick Fox and Robert Kenward who generously allowed me to use their unpublished material. Others who readily responded to my appeals for information include Christian de Coune, Kent Carnie, John Hegan, Christian Saar and John Fairclough. I am grateful to all of them.

REFERENCES

Anon. (1993) Annual Report, The Peregrine Fund, Boise, Idaho.

Berry, R.B. (1972) Reproduction by artificial insemination in captive American goshawks. *J. Wildl. Manage.*, **36**, 1283–88.

Bird, D.M., Burnham, W. and Fyfe, R.W. (1985) A review of cross-fostering in birds of prey, in *Conservation Studies on Raptors* (eds I. Newton and R.D. Chancellor). International Council for Bird Preservation, Cambridge, England, pp. 433–8.

Boyd, L.L. (1978) Artificial insemination of falcons. *Symp. Zool. Soc. Lond.*, **43**, 73–80.

Boyd, L.L. and Schwartz, C.H. (1985) Training imprinted semen donors, in *Falcon Propagation: A Manual on Captive Breeding* (revised edition) (eds J.D. Weaver and T.J. Cade). The Peregrine Fund, Boise, Idaho, pp. 24–31.

Cade, T.J. (1985) Peregrine recovery in the United States, in *Conservation Studies on Raptors* (eds I. Newton and R.D. Chancellor). International Council for Bird Preservation, Cambridge, England, pp. 331–42.

Carson, R. (1962) *Silent Spring*. Houghton Mifflin, Boston.

Cromie, R. and Nicholls, M. (1995) *The Welfare and Conservation Aspects of Keeping Birds of Prey in Captivity*. Report to the RSPCA, London.

De Bastyai, L. (1973) The history of hawking in Hungary, *J. Brit. Falconers' Club*, 6 (2), 114–22.

Durman-Walters, D. (1994) *The Modern Falconer*. Swan Hill Press, England.

Evans, I.M., Love, J.A., Galbraith, C.A. and Pienkowski, M.W. (1994) Population and range restoration of threatened raptors in the United Kingdom, in *Raptor Conservation Today* (eds B-U. Meyburg and R.D Chancellor). WWGBP/The Pica Press, pp. 447–57.

Fox, N.C. (1995) *Understanding the Bird of Prey*. Hancock House, UK.

Grier, J.W., Berry, R.B. and Temple, S.A. (1972) Artificial insemination with imprinted raptors. *J. N. Amer. Falconers' Assoc.*, 11, 45–55.

Gupta, P.K. (1986) *Pesticides in the Indian Environment*. Interprint, New Delhi.

Hamerstrom, F. (1970) *An Eagle to the Sky*. Iowa State University Press, Ames, Iowa.

Hurrell, L.H. (1973) On breeding the sparrowhawk in captivity, in *A Hawk for the Bush* (2nd edition) by J. Mavrogordato. Spearman, London.

Jones, C. (1990) Mauritius kestrel comeback, in *Mauritius Wildlife Appeal Fund Annual Report, 1988–1989*, pp. 9–11.

Jones, C. and Owadally, A.W. (1981) The world's rarest falcon: the Mauritius kestrel. *J. Brit. Falconers' Club*, 7 (5), 322–7.

Kenward, R.E. (1977) The goshawk saga. *J. Brit. Falconers' Club*, 7 (1), 40–6.

Kenward, R.E. (1981) What happens to goshawks trained for falconry? *J. Wildl. Manage.*, 45 (3), 803–6.

Kenward, R.E. (1987) Protection versus management in raptor conservation: the roles of falconry and hunting interests, in *Breeding and Management in Birds of Prey* (ed. P.J. Hill). University of Bristol, pp. 1–13.

Kenward, R.E. (1993) Modelling raptor populations: to ring or radio-tag? in *Marked Individuals in the Study of Bird Population* (eds J.D. Lebreton and P.M. North). Birkhauser Verlag, Basel, pp. 157–67.

Kenward, R.E., Marcström, V. and Karlbom, M. (1991) The goshawk (*Accipiter gentilis*) as predator and renewable resource. *Gibier Faune Sauvage*, 367–78.

Lindberg, P. (1985) Population status, pesticide impact and conservation efforts for the peregrine (*Falco peregrinus*) in Sweden, with some comparative data from Norway and Finland, in *Conservation Studies on Raptors* (eds I. Newton and R.D. Chancellor). International Council for Bird Preservation, Cambridge, England, pp. 343–51.

Lovegrove, R. (1990) *The Kite's Tale: the Story of the Red Kite in Wales*. Royal Society for the Protection of Birds, Bedfordshire.

Monneret, R.J. (1977) Project peregrine, in *Papers on the Veterinary Medicine and Domestic Breeding of Diurnal Birds of Prey* (eds J.E. Cooper and R.E. Kenward). British Falconers' Club, Tamworth, pp. 64–9.

Nelson, R.W. (1977) On the diagnosis and 'cure' of imprinting in falcons which fail to breed in captivity, in *Papers on the Veterinary Medicine and Domestic Breeding of Diurnal Birds of Prey* (eds J.E. Cooper and R.E. Kenward). British Falconers' Club, Tamworth, pp. 39–49.

Newton, I. (1994) Current population levels of diurnal raptors in Britain. *The Raptor*, 21, 17–21. Hawk and Owl Trust, London.

Parry-Jones, J. (1988) *Falconry, Care, Breeding and Conservation.* David and Charles, Newton Abbot.

Saar, C. (1985) The breeding and release of peregrines in West Germany, in *Conservation Studies on Raptors* (eds I. Newton and R.D. Chancellor). International Council for Bird Preservation, Cambridge, England, pp. 363–5.

Sherrod, S.K., Heinrich, W.R., Burnham, W.A., Barclay, J.H. and Cade, T.J. (1982) *Hacking: A Method for Releasing Peregrine Falcons and other Birds of Prey.* The Peregrine Fund, Boise, Idaho.

Speer, G. (1985) Population trends of the peregrine falcon (*Falco peregrinus*) in the Federal Republic of Germany, in *Conservation Studies on Raptors* (eds I. Newton and R.D. Chancellor). International Council for Bird Preservation, Cambridge, England, pp. 359–62.

Treleaven, R.B. (1958) The non-breeding of peregrines in the west of England. *J. Brit. Falconers' Club,* 3 (5), 158–9.

Trommer, G. (1992) Hawk flight arrangement. *J. Brit. Falconers' Club,* 32–33.

USFWS (1988) *Final Environmental Assessment – Falconry and Raptor Propagation Regulations, July 1988.* US Fish and Wildlife Service, Washington, DC.

Weaver, J.D. (1985a) Artificial insemination, in *Falcon Propagation: A Manual on Captive Breeding* (revised edition) (eds J.D. Weaver and T.J. Cade). The Peregrine Fund, Boise, Idaho, pp. 19–22.

Weaver, J.D. (1985b) Imprinted females, in *Falcon Propagation: A Manual on Captive Breeding* (revised edition) (eds J.D. Weaver and T.J. Cade). The Peregrine Fund, Boise, Idaho, pp. 31–3.

Wilson, A.C. and Stanley Price, M.R. (1994) Reintroduction as a reason for captive breeding, in *Creative Conservation: Interactive Management of Wild and Captive Animals* (eds P.J.S. Olney, G.M. Mace and A.T.C. Feistner). Chapman & Hall, London, pp. 243–64.

Woodford, M.H. (1971) *A Manual of Falconry* (2nd edition). A. & C. Black, London.

Wyllie, I. and Newton, I. (1991) Demography of an increasing population of sparrowhawks. *J. Anim. Ecol.,* 60, 749–66.

-10

Deer management in Scotland

Pete Reynolds and Brian Staines

10.1 INTRODUCTION

As a wildlife resource, the red deer (*Cervus elaphus*) population in Scotland is of considerable ecological, utilitarian and aesthetic importance. Scotland's largest native herbivore, red deer, alone or in combination with domestic stock, can drive vegetation successions in the uplands. For example, where browsing is sufficiently intense, woodland is converted to open grass or dwarf-shrub dominated habitats. Therefore, red deer have profound implications for vegetation and associated animal communities.

From a utilitarian perspective, red deer are the primary land-use product from about 1 million hectares (13%) of Scotland's land area (Cooper and Mutch 1979). The 70 000 animals culled annually provide a source of both sport and venison valued at around £1.5 million and £6 million per annum respectively (Scottish Natural Heritage 1994). These activities in turn generate employment in remote and vulnerable rural communities, with over 300 permanent and 458 part-time staff being directly involved in deer management (Callander and MacKenzie 1991). However, the value of red deer as a wildlife resource transcends the purely ecological and utilitarian. The eminent ecologist Frank Fraser Darling wrote:

> These creatures are more than the material of the scientist's paper. They are animals whose lot has been closely linked with human development. We have pitted our wits against them through thousands of years and the hunter is not worth his salt who does not admire this quarry and is not content sometimes to watch the beauty of their lives, free from the desire to kill.
>
> (Darling 1937)

There is, then, an important aesthetic and cultural element. For many people, red deer are the quintessence of the Scottish Highlands.

Conservation and the Use of Wildlife Resources. Edited by M. Bolton.
Published in 1997 by Chapman & Hall. ISBN 0 412 71350 0.

10.2 HISTORICAL PERSPECTIVE

Red deer have been present in Scotland since the end of the last Ice Age, about 11 000 years ago, and are therefore an integral part of the post-glacial upland ecosystems which we see in Scotland today. Archaeological evidence reveals a long history of utilization of the red deer resource by human communities. Excavations of Mesolithic and Neolithic sites, for example, have shown that red deer provided an important source of meat, tools (fashioned from bone and antler) and clothing (Ritchie 1920; Mellars 1987).

The earliest records of red deer being hunted purely for sport date back to the medieval period. Not only were animals coursed with stag hounds, but large-scale deer drives were organized, with kills of over 300 animals per day being recorded (Ritchie 1920). These deer drives continued to the end of the eighteenth century. Together with the loss of forest habitat and competition with sheep, driving probably caused a decline in the red deer population, which reached its nadir in the mid to late 1700s. Throughout this time the only form of management was primarily legislative in nature. Various Acts defined the means by which deer could be 'taken', and by whom, and the nature of penalties.

By the early 1800s specific areas of land ('Deer Forests') were being set aside to encourage deer populations. Landowners were motivated by the commercial prospects of sporting revenues. Stalking (the pursuit of deer, principally stags, on foot) was becoming fashionable among the wealthy as a consequence of improved firearm technology and improved rail communications with England. The number of Deer Forests increased from 45 in 1838 to 203 in 1912, covering about 1.5 million hectares or approximately 19% of the land area of Scotland (Darling 1955). The seeds of contemporary deer management were sown.

10.3 CONTEMPORARY PERSPECTIVE

10.3.1 Numbers and distribution

Attempts to estimate the total population of red deer in Scotland have, until recently, been little more than intelligent guesses. More objective estimates became possible in 1959 following the introduction of the Deer (Scotland) Act, which, among other things, required the Secretary of State to constitute a Red Deer Commission (RDC) with the general functions of furthering the conservation and control of red deer. Since 1959 the RDC has systematically counted large areas of Scotland annually. However, different areas have been counted at irregular intervals and in consequence the estimation of changes in total numbers, other than in crude terms, is difficult. Clutton-Brock and Albon (1989) attempted to overcome this problem by using a multiple regression model to obtain total population estimates. Their results

Table 10.1 Estimates of the total open hill red deer population in Scotland, 1900–1986

Year	Estimate	Source
1900+	150 000	Cameron (1923)
1912	180 000	Darling (1955)
1930	250 000	Parnell (1932)
1960–1965	216 000 ± 45 000	Clutton-Brock and Albon (1989)
1966–1969	155 000 ± 43 000	Clutton-Brock and Albon (1989)
1975	248 000 ± 55 000	Clutton-Brock and Albon (1989)
1986	297 000 ± 40 000	Clutton-Brock and Albon (1989)

are included in Table 10.1. The estimates shown in Table 10.1 refer to open-hill deer. If deer in forestry are included, the most recent estimates should probably be increased by a further 25–50 000 (Stewart 1986; Staines and Ratcliffe 1987). The contemporary resource thus comprises in excess of 300 000 animals ranging over some 3 million hectares or more than 40% of Scotland's land area.

Overall deer densities vary from <5 to >20 per km^2 (Table 10.2) and appear to be related to variations in environmental quality, culling and colonization rates and the extent of competition with sheep (Clutton-Brock and Albon 1989).

Table 10.2 Range of red deer densities found on open hill land in Scotland. From Staines and Ratcliffe (1987) and Mitchell and Crisp (1981)

Count area	Density (deer/100 ha)
Rovie/Skibo	0.5
Skye	1.6
North Sutherland	4.3
North Ross	4.8
South Ross	13.1
West Grampians	13.9
East Grampians	14.1
Glenartney	26.9
Scalpay	31.3
Scarba	34.4

10.3.2   Population attributes

The Scottish red deer population has almost doubled in the last 30 years (Table 10.1). Sequences of dry summers and mild winters, declines in sheep numbers and under-culling of hinds have all been implicated as possible causes (Clutton-Brock and Albon 1989). Within areas counted twice by the RDC, increases have ranged from 11% to 150%.

The contemporary population is also characterized by a female bias in the adult sex ratio. Data obtained from the 39 RDC count areas between 1983 and 1994 show a mean ratio of 1.9 females per male ($n = 39$, sd = 0.43, range = 1.2–3) (data derived from Appendix A, RDC Annual Report for 1993/94).

Such bias is not unusual in ungulate populations which increase to a point where they are naturally regulated by the availability of resources. Under these circumstances, both juvenile and adult males suffer higher mortality rates than females, due to the higher energy demands associated with larger body size and thus greater susceptibility to starvation (Clutton-Brock and Albon 1989; Clutton-Brock 1991). Within Scotland males are frequently subjected to higher culling rates than females and this will also tend to favour a female bias in the population.

Originally an animal of open woodland and woodland edge, about 80% of the Scottish red deer population now inhabit the extensive open and relatively infertile habitats characteristic of the Scottish uplands. This adaptation has been associated with a reduction in body size, growth rates and reproductive output relative to resident woodland red deer. For example, the influence of habitat is reflected in potential birth rates of around 43 and 65 calves per 100 hinds aged 1 year and above for open hill and woodland deer respectively (Table 10.3). A birth rate of about 70 calves/100 hinds is believed to be the maximum achievable by red deer, necessitating pregnancy as yearlings (i.e. first calving as 2-year-olds) and

Table 10.3 Potential birth rates of red deer in woodland and open habitats in Scotland

Calves/100 hinds	Habitat	Source
70	Woodland	Mitchell *et al.* (1981)
69–70	Woodland	Ratcliffe (1984a)
30–50	Open hill	Clutton-Brock and Albon (1989)
44	Open hill	Mitchell *et al.* (1986)
46–47	Open hill	Mitchell and Crisp (1981)
38–43	Open hill	Lowe (1971)

breeding every year thereafter. Corresponding birth rates for hinds becoming pregnant as 2- and 3-year-olds would be 60/100 and 50/100 respectively. In practice these maximum rates are rarely achieved in open-hill populations, although birth rates of 60–70 are common in many forest populations (Ratcliffe 1984b). The comparatively low birth rates of Scottish open-hill populations can be explained by retarded growth, the associated late attainment of puberty and poor breeding success of adult hinds. Mitchell (1973) and Mitchell and Brown (1974) showed that a majority of open-hill hinds attain puberty at 2 years old, calving for the first time at age 3, although a few yearlings (mean of 7% from all areas) became fertile in some parts of Scotland.

A proportion of adult hinds only produce calves in alternate years due to the high nutritional demands associated with pregnancy and lactation in the previous year (Mitchell 1973). These so-called 'yeld' hinds comprised 34% and 45% of the adult hind populations at two different sites studied by Mitchell *et al.* (1986) and Mitchell and Crisp (1981) respectively.

From a resource management point of view, the number of new animals recruiting into the population each year determines the number that can be harvested from the population. For open-hill red deer populations the average level of natural calf mortality is about 20–25% (RDC 1981), with half of this occurring during and immediately after the calving season and the rest during late winter to early spring. An average birth rate of 43 calves per 100 hinds therefore results in a spring recruitment rate of about 34 surviving calves per 100 hinds following a reasonable winter. In particularly favourable winters calf mortality may be as low as 10%, increasing up to 60–75% under severe conditions. Further details are provided in Table 10.4.

Most natural mortality occurs in late winter/early spring, and is highest amongst calves (and to a lesser extent yearlings) and the older age classes (Mitchell *et al.* 1986; Clutton-Brock and Albon 1989). As in many other mammals, high population density and low food availability affect the survival of males more than females. Clutton-Brock and Albon (1989), for example, found that as population density increased in an unculled population on the island of Rum in western Scotland, the mortality of males during their first 2 years of life rose from less than 10% to around 60%, whilst the mortality of females over the same period increased from about 10% to 30%. Following these initially high rates, natural mortality then remains relatively low up to about 8–10 years of age in both hinds and stags, after which rates once more increase (Mitchell *et al.* 1977, 1986; Mitchell and Crisp 1981; Clutton-Brock and Albon 1989).

Data on natural mortality rates for adult animals (i.e. excluding calves) are summarized in Table 10.5. These figures were derived from managed populations which are subjected to culling during the period July to October (stags) and October to February (hinds). Most natural mortality

Table 10.4 Calf recruitment and natural mortality rates (from birth to following spring). All counts undertaken by staff of the Red Deer Commission

Calf/hind ratio at birth	Calf/hind ratio in following spring	Calf mortality (birth to following spring)	Source
44/100	35/100	20%	Mitchell *et al.* (1986)
46/100	30/100	35%	Mitchell and Crisp (1981)
–	–	23%*	McLean (1993)
–	33/100[†]	–	Red Deer Commission (1993)
–	30–50/100	–	Clutton-Brock and Albon (1989)
–	–	9.5%	Lowe (1969)
61/100	55/100[§]	*c* 10%	Ratcliffe (1987)
–	–	14%**	Clutton-Brock *et al.* (1995)

* Average figure derived from 16 estates in 1993.
† Average figure derived from counts in 39 Red Deer Commission count areas undertaken between 1983 and 1994 ($n = 39$, sd = 0.07, min. = 0.19, max. = 0.46).
§ These figures are from a woodland population.
** Mean figure for Rum, 1981–1994 ($n = 52$, sd = 13.37, min. = 0, max. = 67.6).

Table 10.5 Adult mortality data

Stags	Hinds	Stag + hinds	Source
3.5%	4.3%	3.9%	*1
1.9%	3.2%	2.6%	*2
5.6%	5.8%	5.7%	*3
3.9%	4.1%	–	*4

*1, Data derived from Table 18 of Lowe (1969) for period 1958–1965 from island of Rum. Mean figure shown (stags: $n = 8$, sd = 2.04, min = 1.3%, max. = 7%, hinds: $n = 8$, sd = 3.03, min. = 2%, max. = 11.5%; stags and hinds combined: $n = 16$, sd = 2.53, min. = 1.3, max. = 11.5).
*2, Data derived from Tables 5 and 7 of Mitchell *et al.* (1986) for Glenfeshie. Mean figure shown (stags: $n = 8$, sd = 0.86, min. = 1%, max. = 3.2%; hinds: $n = 8$, sd = 1.54, min. = 1%, max. = 5.3%; stags and hinds: $n = 16$, sd = 1.01, min. = 1%, max. = 5.3%).
*3, Data derived from Mitchell and Crisp (1981) for island of Scarba based on 1974 population of 122 calves, 234 stags and 293 hinds (figure shown not an average).
*4, Data from McLean (1993). Mean figures for 16 estates in the Central Highlands for 1993.

occurs during late winter/early spring, i.e. following culling. In terms of longevity, Mitchell *et al.* (1986) found that only 13 (0.4%) of 3440 deer shot or found dead at Glenfeshie were 16 or more years old, with the oldest animals being 20 (stag) and 21 (hind). Elsewhere red deer have been shown to live to a maximum of about 20 years (Mitchell 1970). Most aspects of the growth, reproduction and survival of red deer decline in response to increasing population density. In a classic study on the island of Rum, a sub-population of deer was allowed to increase from a density of about 18 deer per km^2 to about 25 deer per km^2 over the period 1971–1991 (see Clutton-Brock and Albon 1989). As density increased:

- age of hinds at first breeding increased, with the percentage of 2-year-olds conceiving falling from 65% to 10%
- proportion of milk hinds calving declined from 90% to 30%
- overwinter mortality of calves increased from <5% to 40% and of year-lings from 0 to 30%. This effect was particularly pronounced in males, mortality in the first two years of life increasing from <10% to *c.* 60%, compared with 10%–30% for females over the same period
- body and antler growth of males declined, antler weights decreasing by 20%

During the period 1972–1991 hinds increased from 56 to 160 while stags initially increased from 116 to 155 in 1978 and then declined to 100. Adult sex ratios over the same period changed from 1 stag : 0.5 hinds to 1 stag : 1.6 hinds (Clutton-Brock and Lonergan 1994). The population thus became female-biased. That such density-dependent effects are not confined to the Rum study area is suggested by comparison with data from additional sites in Scotland covering a range of densities from 1.6 to 34.4 deer per km^2 (e.g. Mitchell *et al.* 1977, 1986; Staines 1978; Mitchell and Crisp 1981; Albon *et al.* 1983). The existence of these density-dependent influences on growth, reproduction and survival has profound implications for those managing the wild deer resource.

Males in particular seem to be disadvantaged by high population density. Given their faster growth rates, larger size and thus higher metabolic requirements, males are more susceptible to food shortage than females. Based on evidence that stags are excluded from preferred habitat by rising hind numbers (Staines *et al.* 1982; Clutton-Brock *et al.* 1987), Clutton-Brock *et al.* (1985) speculated that stag performance may be more strongly affected by hind numbers than by stag density. For example, in the Rum study areas the best predictors of stag growth and survival were hind, not stag, densities (Clutton-Brock and Albon 1989). This hypothesis is currently being tested in a series of population manipulations on Rum. If correct, then aspects of contemporary deer management in Scotland may be shown to be inappropriate and counter-productive.

10.3.3   Population management

The dominant activity associated with the management of red deer in Scotland is culling. The number of animals shot annually has increased from approximately 24 000 animals in 1973 to 70 000 in 1992/93 and 57 300 in 1993/94 (RDC 1993). Of these (data from Callander and MacKenzie 1991):

- 25% are culled by forestry and agricultural interests to reduce or prevent damage to crops
- 50% are shot by estate staff for crop protection and for the purposes of 'improving' the sporting resource. This latter activity is sometimes referred to as the management cull
- 25% are shot for sport

In most parts of the Highlands 6–12% of the hinds and 10–17% of stags are killed each year, these culling rates being calculated as the percentage of the spring population, excluding calves, killed in the subsequent autumn/winter (Clutton-Brock and Albon 1989). More recent figures are shown in Table 10.6.

Based on the highest potential recruitment rate of about 55 calves per 100 hinds (Table 10.4), the maximum sustainable cropping rate is some 27% of adults annually, assuming equal adult sex ratios. However, for most red deer populations in the barren uplands of Scotland, average recruitment rates are much lower, being in the region of 33 calves/100 hinds. As a consequence, the maximum sustainable cropping rate is about 16% of the adults annually (again assuming equal adult sex ratios) and accordingly a one-sixth cull has been advocated by the RDC (RDC Annual Reports 1961–1975).

Clutton-Brock and Albon (1989) concluded that, throughout much of

Table 10.6 Average culling rates (%) for 17 of the Red Deer Commission count areas, covering the period 1988–1993

	Males	Females
Mean	17.9	15.2
Standard deviation	9.7	6.3
Min.	9	4.6
Max.	53.3	27.9

Culling rate calculated as the % of the spring population, excluding calves, killed in the subsequent autumn/winter. Raw data from Appendix A (Red Deer Commission 1993).

Scotland, the number of hinds culled was lower than the rate of recruitment. Thus, in 1986, more than 75% of the 42 RDC count areas into which Scotland is divided were culling fewer hinds than the number of animals recruited and over 70% were culling stags at or above the rate of recruitment. More recently there is evidence that, at least for hinds, culling and recruitment rates are approaching equilibrium (Table 10.6).

Most culling is undertaken by stalking, involving locating animals on foot and attaining a position from which the quarry can be humanely shot. The sporting cull, and to a lesser extent the management cull, are highly selective. The sporting cull is directed at stags between 7 and 10 years old, with antlers of good size and shape (i.e. high trophy value) or at stags considered to have unfavourable characteristics. The latter animals are removed in the hope of improving the quality of stock. Although there is some evidence that antler shape may be genetically determined (Ahlen 1965), the role of environment would seem to be more important in influencing antler development. For example, Clutton-Brock and Albon (1989) showed that both antler weight and antler length were negatively correlated with hind population density. Selective shooting may therefore have little effect on trophy quality.

While the trophy is perhaps becoming secondary to the challenge and experience of stalking mature stags in difficult terrain, contemporary deer management appears reluctant, in some respects, to relinquish the Victorian preoccupation with 'heads'. The consequence has been a shortage of mature stags for stalking with an increased culling of younger stags. The age distribution of the cull appears to have shifted from stags that were on average about 9 years old and weighing 92 kg, to stags that are much younger at 4–5 years old, weighing an average of 80 kg and with inferior heads (RDC 1990) weights given are 'larder' weights, i.e. complete animal less blood and alimentary tract).

That sportsmen prefer stalking stags is reflected in the average fee paid for shooting an open-hill stag of about £200–£250, compared with £50 for hinds. With about 31% of stags being shot on a commercial basis, total income from these commercial lets is probably in the region of £1.5–£2 million (Callander and MacKenzie 1991). On many of the approximately 500 privately owned estates which manage red deer, the number of stags culled annually appears to be determined by tradition, rather than by population dynamics, with a fixed number or fixed proportion being taken each year.

Less than 5% of hinds are shot for sport, and less than half of these on a commercial basis with a current fee of around £50 per hind (Callander and MacKenzie 1991). Most are culled by estate staff for the purposes of stock management, although which category of animal should be culled, other than the sick or injured, is a matter of conjecture. Traditionally, lactating females ('milk hinds') and their calves were not shot and culling was therefore concentrated on the non-breeding females. Given that about 30% of these may be immature and many of the remainder will be pregnant, it has been argued that this policy removes valuable potential breeding stock (RDC 1981). However, such a traditional cull does produce better quality venison, since non-breeding hinds generally enter the winter in better condition than milk hinds. Contemporary culls appear to include an increasing proportion of milk hinds. For example, assuming all hinds were culled with their calves, then the proportion of milk hinds in the cull was 18.8% in 1985/86 and 28.1 in 1990/91 (RDC Annual Reports).

The annual hind cull, as for stags, appears to be determined arbitrarily on many estates and to be dictated by tradition rather than by population dynamics. A similar arbitrary approach prevails with respect to the definition of the optimum size of sub-populations of red deer. The RDC is now beginning to address this problem by defining, for each of the approximately 50 Deer Management Group Areas within the deer range, the minimum populations necessary to sustain traditional stag culls, based on annual recruitment of 32% and a culling rate of 16%. Nevertheless, this approach reflects the extent to which contemporary management of red deer range continues to be animal and stalking estate orientated. Deer, rather than the soils and vegetation, are regarded as the resource. Rather than posing the question how many deer are required to support a given stag cull we should, in the first instance, be posing the question as to how many deer the range can support. Although the term 'carrying capacity' is fraught with difficulties in terms of precise definition, it is nevertheless a useful concept and implies range-induced constraints on population growth and productivity (e.g. see Nicholson 1974; Macnab 1985). Given the relatively poor performance of open-hill deer in Scotland as a consequence of density-dependent suppression of growth and fecundity, it can be argued that many populations are at or near carrying capacity. That range condition and range capacity figure little in contemporary population management is lamentable. Not only is the attainment of maximum sustainable yields of trophy stags and/or venison compromised, but as a consequence the red deer population in Scotland has become a source of conflict with other land uses.

10.3.4   Resource management problems

Although the total annual cull has increased substantially, the cull rate has lagged behind recruitment rate, resulting in a doubling of the population within the last 30 years. The under-culling of hinds is likely to have been particularly relevant in this respect, the resulting female-biased populations being highly productive (at the population level) and demanding high culling rates to achieve stability (Table 10.7). This increasing population of red deer in Scotland has been the source of much conflict of interests and as

Table 10.7 The effect of adult sex ratio on the culling rate required to maintain a stable population of red deer

Sex ratio stags : hinds	Recruitment rate	Culling rate (%)
1 : 1	0.35	17.5
1 : 2	0.35	23.3
1 : 3	0.35	26.1

*Recruitment rate refers to the number of calves per 100 hinds in spring.
†Culling rate refers to the number of animals removed in autumn/winter as a proportion of the total number of animals 1 or more years old in the preceding spring.

a consequence the traditional approach to deer management is being questioned (e.g. Callander and McKenzie 1991; Scottish Natural Heritage 1994).

The agricultural protection cull averages some 3500 deer per annum, or about 7% of the total red deer cull. As assessed by the proportion of stags and hinds killed out of season, stags appear to be the more likely to maraud agricultural crops (Table 10.8). That stags should predominate in the agricultural protection cull may, to some extent, reflect the combined effects of their relatively high energy requirements and high hind densities. In the latter situation, swards tend to be closely cropped, favouring the smaller-mouthed females. In contrast, stags, with their larger mouths, appear to be disadvantaged on such closely cropped swards and, in order to maintain their intake rate, may be forced to feed in areas where food availability is relatively high (Watson and Staines 1978; Staines *et al.* 1982; Clutton-Brock and Albon 1989). Marauding of agricultural crops and the

Table 10.8 Proportion of stags, hinds and calves culled because they were marauding agricultural crops

	Stags	Hinds	Calves
Mean	81.3%	16.2%	2.6%
Standard deviation	9.13	7.73	2.17
Min.	68%	6%	0.3%
Max.	93%	27%	6%

Data derived from Table 4 of Red Deer Commission Annual Report for 1993/94, based on authorizations issued to estates and farms for out-of-season shooting of marauding deer for each of the years 1984–1994. Annual number culled ranged from 172 to 562 animals: $n = 10$.

shooting of the offending animals may be a consequence. For the manager of traditional sporting estates, this loss of potential trophy animals is a source of frustration.

From the forester's perspective, conflicts arise from the damage caused by red deer as a consequence of browsing and bark-stripping. The numbers of deer culled for the purposes of forest protection have increased from about 5000 in 1976 to around 7000 in 1993/94, equivalent to some 12% of the annual cull (RDC Annual Reports for 1976 and 1993/94). In contrast to the agricultural cull, the animals killed for forest protection are more equitably distributed between the sexes (Table 10.9).

Large sums of money are spent in attempts to control forest deer populations. While these costs are to some extent offset by income generated from venison, stalking and trophy charges, the cost of the protection cull is considered to greatly exceed income from shooting in all Scottish forests (RDC 1990). The problem of deer damage to forestry is exacerbated by the high productivity and low mortality characteristic of woodland populations (Ratcliffe 1984b) and the consequent need for culling rates in excess of 18–25%. A further problem arises from the fact that the design of many of our plantation forests is not conducive to effective control of red deer populations, with dense plantings and few open areas inhibiting shooting. In addition the relatively short 35- to 55-year rotations which characterize commercial forests in Scotland result in relatively high proportions of pre-thicket- and thicket-stage plantation being available. These are precisely the growth stages at which red deer densities are at or near their maximum (Ratcliffe *et al.* 1986; Staines and Ratcliffe 1987).

Through their grazing, browsing and trampling, the increasing population of red deer in Scotland is also one of the potential agents of observed vegetation change in the uplands which is causing widespread concern. (e.g. Callander and MacKenzie 1991; Scottish Natural Heritage 1994). However, because red deer share much of their range with some 2 million sheep, it is

Table 10.9 Mean proportion of stags, hinds and calves culled for forest protection 1986/87–1993/94

	Stags	Hinds	Calves
Mean	46.8%	39%	14%
Standard deviation	2.25	2.67	2.1
Min.	43%	37%	12%
Max.	50%	45%	18%

Data derived from Table 11 of Red Deer Commission Annual Report for 1993/94. Annual number culled ranged from 5312 to 7744: $n = 8$.

not always clear to what extent the observed changes are attributable to which species. The demise of our native woodlands, formerly widespread but now covering less than 2% of Scotland, is an exception in this respect.

The loss of native woodland dominates the contemporary debate on deer and the natural heritage. In most of the remaining woodlands little or no regeneration is taking place due to browsing, and in many cases deer are primarily responsible. The demise of our remaining native Caledonian pinewoods is typical of the problem with which we are confronted if we wish to conserve these habitats for their own sake and for the maintenance of biodiversity and soil productivity. Studies by Steven and Carlisle (1959) and Watson (1983) suggest that in many of these woodlands little or no regeneration has taken place, in some cases for over 200 years (when the Scottish red deer population was probably at its nadir). The trees may be reaching the end of their seed-bearing lives and reductions in grazing pressure are required if regeneration is to be promoted.

Attempts have been made to define winter densities of deer at or below which tree regeneration will take place (Table 10.10). Densities of about 5 to 6 deer per km^2 appear to be critical, although our knowledge of the precise relationship between deer density and tree regeneration remains inadequate. On the basis of these figures, contemporary total deer densities are incompatible with tree regeneration in 83% of the 29 RDC count areas for which data are available (1980–present). However, the factors influencing the impact of red deer on tree regeneration are complex and the simplistic compatible densities given above must be seen in this context. Woodlands in the uplands can benefit from light grazing by large herbivores

Table 10.10 Response of tree regeneration to red deer density

Winter deer density (deer/km^2)	Species	Response	Source
33–50	Scots pine *Pinus sylvestris* Birch *Betula pendula*	Regeneration prevented	Cummins and Miller (1982)
25	Scots pine *Pinus sylvestris*	Regeneration prevented	Holloway (1967)
6	Various	Some regeneration	Darling (1937)
4	Scots pine *Pinus sylvestris*	Some regeneration	Holloway (1967)
2	Scots pine *Pinus sylvestris*	Regeneration	Holloway (1967)

such as red deer. These animals produce and maintain structural diversity, both in terms of plant communities and dependent woodland faunas. In addition, ground disturbance by grazing animals creates niches for tree seedling establishment (Mitchell and Kirby 1990; Sykes 1992). The ultimate goal, therefore, should be the regeneration and expansion of our native woodlands in the presence of red deer, albeit at reduced densities.

Increasing numbers of red deer have also been implicated in the decline of heather moorland, dominated by *Calluna vulgaris*, in the Scottish uplands. Within the 12 Scottish Regions, net losses of this habitat averaged 23% (range 2–67%) between the 1940s and 1970s, due largely to afforestation and to succession from heather moorland to unimproved grassland as a consequence of heavy grazing (Thompson *et al.* 1995). However, teasing out the relative roles of red deer and sheep in this latter respect is difficult. As a consequence, evidence for the role of red deer as agents of widespread heather loss is inconclusive, although they are clearly capable of causing localized losses (e.g. Watson 1989; Welch *et al.* 1992).

More generally, concern has been expressed that red deer, through their browsing and trampling, may be contributing to an overall deterioration in the condition of upland habitats, for example by promoting soil erosion and preventing the development of altitudinal continuity in near-natural vegetation from woodland through scrub and dwarf-shrub heaths to montane summits (Francis *et al.* 1991). However, objective data concerning the role of red deer in promoting damage to, or loss of, plant communities other than native woodland are scarce.

10.4   TOWARDS INTEGRATED RESOURCE MANAGEMENT

The problems posed by an increasing red deer population have festered since the second half of the 1800s. For example, during the period 1872–1954 there were seven government appointed enquiries concerning red deer, three of which considered damage to crops. The Deer (Scotland) Act 1959 contained provisions for the control of deer for the purpose of preventing damage to agriculture and woodland and established the RDC with overall responsibility for the conservation and control of wild red deer in Scotland.

Throughout its existence the RDC has advocated a reduction in the population of red deer. However, the population has doubled and the concerns of natural heritage have been added to those of agriculture and forestry.

10.4.1   Population dynamics models

Sporting estates frequently seek to maximize the number and quality of mature stags that can be harvested each year. Due to competition, increasing population density depresses the growth and survival of males more than of females and as a consequence different culling policies may be required for the two sexes. The potential advantages of the use of population models, incorporating density dependence, for predicting maximum sustainable yields and optimum population size are illustrated by two recent publications.

Clutton-Brock and Lonergan (1994) quantified observed density-dependent changes in reproduction and survival over a period of 21 years (1970–1991) in an unculled population of red deer on the island of Rum. They used these relationships to model the responses of males and females to different levels of culling and showed that yields of venison or stags could be increased by reducing population density, especially of hinds. The maximum yield of mature stags (5 or more years old) was achieved at hind culling rates of 16–20%. Clutton-Brock and Lonergan concluded that those estates culling less than 5% of their hinds may be able to increase their annual offtake of mature stags by as much as 30% if the 16–20% hind culls were to be adopted.

More recently, Buckland *et al.* (1996) have also developed a population dynamics model which they used to determine the minimum populations necessary to sustain specific culls of mature stags (6 or more years old). Their model is based on data from four sites in Scotland, including Rum. The authors conclude that due to density-dependent effects on hind fertility and calf survival, many estates may be able to sustain large reductions in numbers, yet still maintain the stalking value of the deer population. It is argued that, for populations managed at densities well below carrying capacity of the range, for every 10 mature stags to be stalked annually the post-cull population need only comprise 60 stags, 30 hinds and 20 calves (total of 110).

Contemporary deer-cull ''rules of thumb' (which do not take into account the effects of density-dependence) are based on the one-sixth cull advocated by Lowe (1969) and are used by the RDC in providing advice on populations required to sustain stag culls. Using this convention, for every 10 mature stags to be stalked annually, a total population of 220 would be required (assuming an adult sex ratio of 1 stag : 2 hinds) or 164 (assuming an adult sex ratio of 1 stag : 1.3 hinds).

The implications for contemporary deer managers of the modelling work undertaken by Buckland *et al.* and Clutton-Brock and Lonergan are radical. Both models suggest that hind populations could be considerably reduced whilst maintaining stag culls. In addition, stag quality, in terms of size and weight, is likely to improve and Buckland *et al.* speculate that 4-year-olds could conceivably have a trophy value comparable with that of 6-year-olds from under-culled populations in which growth rates and condition are depressed.

Both models also cast doubt on the traditional view that hinds should predominate in a population. In unculled or lightly culled populations hinds do indeed predominate due to high stag mortality. If stag production is the principal management objective, then sex ratios nearer to parity or skewed in favour of males are likely to be beneficial from an economic point of view.

10.4.2   Grazing models

As we have seen, the management of red deer in Scotland is primarily animal, rather than range, orientated. Desirable population size is determined primarily by the number of stags required for sporting purposes. The capacity of the range to support such populations is rarely questioned and yet, over much of the range, it would appear that populations are at or very near carrying capacity. Under these circumstances not only is the productivity and survival of individual animals suppressed due to density-dependent effects, but the risk of long-term habitat change is increased.

We need, therefore, to move towards a situation in which range condition and carrying capacity become fundamental to the management of red deer populations. In order to achieve this we need to be able to estimate the consequences of red deer population density on the utilization of semi-natural vegetation and vegetation dynamics. Given specific management objectives for vegetation it should then be possible to predict how many deer an area could support.

In a collaborative venture several organizations are now developing a computer-based decision support model to estimate the consequences of red deer population size on the use and dynamics of vegetation communities within discrete management units (Gordon 1993). The model will also provide estimates of food intake, and, when linked with a population dynamics model, should provide a much needed objective basis for red deer management.

10.4.3   Management plans

Red deer both influence, and are influenced by, aspects of the range which they occupy. Thus, depending on their density, deer can drive vegetation dynamics such that woodland can be converted to open moorland, with a reciprocal decline in population density and productivity. Deer and range dynamics are inextricably linked and the key characteristics of this range must be taken into account in the determination of optimum deer populations compatible with range management objectives.

The population dynamics and grazing models will need to be used within the context of integrated management plans. These will provide a framework for managing deer as part of a wider whole, rather than in isolation from other interests. In defining optimum deer populations, stalking objectives should be considered *together* with those of other key stakeholders, including agriculture, forestry, natural heritage and recreation, since they all interact in one way or another with deer. Tentative steps towards the production of such plans for discrete deer management areas have been taken by the RDC. The success of these plans is crucial if the conflicts which dominate contemporary deer management in Scotland are to be resolved.

10.5 CONCLUSIONS

Expressions of concern at the manner in which we manage our red deer populations in Scotland have been voiced for over 70 years. Thus, Cameron (1923) wrote:

> Scottish deer stalking wants a new start, and will get it if the rising generation, the schoolboys of today, will insist on making the sport less ceremonious, and will shake it free from its prescriptive leading strings.

Cameron was referring, of course, to the Victorian preoccupation with the stalking of trophy stags and the associated accoutrements, including supplementary feeding and the release of stags from other parts of Europe or even North America with a view to 'improving' the indigenous breeding stock. In particular, the under-culling of hinds has encouraged populations to increase. The associated suppression of growth and fecundity has compromised the attainment of maximum sustainable yields of venison and trophy stags for stalking. Perversely, traditional estate management has been instrumental in eroding the very resource it has endeavoured to encourage.

Victorian preoccupations are steadily being eroded, and slowly a more integrated approach to management is evolving. Fundamental to progress in this respect is the need to move towards range-orientated, rather than traditional animal-orientated approaches to wildlife resource management. Although this will require a reduction in the deer population in many parts of Scotland, the benefits of such a cultural change will be reflected in enhanced habitat condition and in corresponding improvements in deer condition, growth and fecundity. If such a range-orientated and integrated approach to management is adopted, then the ecological, socio-economic and cultural benefits which we derive from the red deer resource will be sustained for future generations. It is appropriate to end with a quote from Cameron, whose foresight has perhaps finally seen the light of day:

> ... we must concentrate our efforts upon improving by every natural means in our power the feeding capacity of our deer forests, and lastly, we must cut down the numbers of our deer till the ground can maintain them without artificial aid.
>
> Cameron (1923)

REFERENCES

Ahlen, I. (1965) Studies of the red deer *Cervus elaphus* in Scandinavia. III, Ecological investigations. *Viltrevy*, 3, 177–376.

Albon, S.D., Mitchell, B. and Staines B.W. (1983) Fertility and body weight in female red deer: a density dependent relationship. *Journal of Animal Ecology*, 52, 969–80.

Buckland, S.T., Ahmadi, S., Staines, B.W., Gordon, I.J. and Youngson, R.W. (1996) Estimating the minimum population size that allows a given annual number of mature red deer stags to be culled sustainably. *Journal of Applied Ecology*, 33, 118–30.

Callander, R.F. and MacKenzie, N.A. (1991) *The Management of Wild Red Deer in Scotland*, Rural Forum, Scotland.

Cameron, A.G. (1923) *The Wild Red Deer of Scotland*. Antony Atha Publishers, Norfolk.

Clutton-Brock, T., Thompson, D. and Covey, C. (1995) Monitoring of Red Deer Changes on Rum. Report to Scottish Natural Heritage, Contract NCCS 024/92/ UPB 1994.

Clutton-Brock, T.H. and Albon, S.D. (1989) *Red Deer in the Highlands*. BSP Professional Books, Oxford.

Clutton-Brock, T.H. and Lonergan, M.E. (1994) Culling regimes and sex ratio biases in Highland red deer. *Journal of Applied Ecology*, 31, 521–7.

Clutton-Brock, T.H. (1991). Sport and the wise use of ungulate populations. *Gibier Faune Sauvage*, 8, 309–17.

Clutton-Brock, T.H., Iason, G.R. and Guinness, F.E. (1987) Sexual segregation and density related changes in habitat use in male and female red deer *Cervus elaphus*. *Journal of Zoology London*, 211, 275–89.

Clutton-Brock, T.H., Major, M. and Guinness, F.E. (1985) Population regulation in male and female red deer. *Journal of Animal Ecology*, 54, 831–46.

Cooper, A.B. and Mutch, W.E.S. (1979) The management of red deer in plantations, in *The Ecology of Even-Aged Forest Plantations*, (eds E.D. Ford and D.C. Malcolm). Institute of Terrestrial Ecology, Cambridge, pp. 453–62.

Cummins, R.P. and Miller, G.R. (1982) Damage by red deer *Cervus elaphus* enclosed in planted woodland. *Scottish Forestry*, 36, 1–8.

Darling, F.F. (1937) *A Herd of Red Deer*. Oxford University Press, London.

Darling, F.F. (1955) *West Highland Survey. An Essay in Human Ecology*. Oxford University Press, Oxford.

Francis, J.M., Balharry, R. and Thomson, D.B. (1991) The implications for upland management: a summary paper, in *Deer, Mountains and Man* (ed. H. Rose). British Deer Society and Red Deer Commission, Inverness, pp. 12–14.

Gordon, I.J. (1993) The development of a decision support system for managing the impact of red deer on vegetation dynamics and habitat diversity. Appendix C, *Red Deer Commission Annual Report 1993/94*, HMSO, Edinburgh.

Holloway, C.W. (1967) The effect of red deer and other animals on naturally regenerated Scots Pine, PhD thesis, University of Aberdeen.

Lowe, V.P.W. (1969) Population dynamics of the red deer *Cervus elaphus* on Rhum. *Journal of Animal Ecology*, 38, 425–57.

Lowe, V.P.W. (1971) Some effects of a change in estate management on a deer population, in *The Scientific Management of Animal and Plant Communities for*

Conservation (eds E. Duffey and A.S. Watt). Blackwell Scientific Publications, Oxford, pp. 437–56.

Macnab, J. (1985) Carrying capacity and related slippery shibboleths. *Wildlife Society Bulletin*, **13**, 403–10.

McLean, C. (1993) Mortality survey. Appendix B, Red Deer Commission Annual Report 1993/94. HMSO, Edinburgh.

Mellars, P. (1987) *Excavations on Oronsay: prehistoric Human Ecology on a Small Island.* Edinburgh University Press, Edinburgh.

Mitchell, B. and Brown, D. (1974) The effect of age and body size on fertility in female red deer *Cervus elaphus. Proceedings of the International Congress of Game Biologists*, **11**, 89–98.

Mitchell, B. (1970) Notes on two old red deer. *Deer*, **2**, 568–70.

Mitchell, B. (1973) The reproductive performance of wild Scottish red deer *Cervus elaphus. J. Reprod. Fert.*, Suppl. 19, 271–85.

Mitchell, B. and Crisp, J.M. (1981) Some properties of red deer (*Cervus elaphus*) at exceptionally high population densities in Scotland. *J. Zool. Lond.*, **193**, 157–69.

Mitchell, B., Grant, W. and Cubby, J. (1981) Notes on the performance of red deer *Cervus elaphus* in a woodland habitat. *Journal of Zoology London*, **194**, 279–84.

Mitchell, B., McCowan, D. and Parish, T. (1986) Performance and population dynamics in relation to the management of red deer *Cervus elaphus* at Glenfeshie, Inverness-shire, Scotland. *Biological Conservation*, **37**, 237–67.

Mitchell, B., Staines, B.W. and Welch, D. (1977) *Ecology of Red Deer: A Research Review Relevant to their Management in Scotland.* Institute of Terrestrial Ecology, Cambridge.

Mitchell, F.J.G. and Kirby, K.J. (1990) The impact of large herbivores on the conservation of semi-natural woods in the British Uplands. *Forestry*, **63**(4), 333–53.

Nicholson, I.A. (1974) Red deer range and problems of carrying capacity in the Scottish Highlands. *Mammal Review*, **4**(3), 103–18.

Parnell, I.W. (1932) Some notes on the natural history of red deer. *Proceedings of the Royal Philosophical Society, Edinburgh*, **22**, 75–80.

Ratcliffe, P. (1984a) Population dynamics of red deer *Cervus elaphus* in Scottish commercial forests. *Proceedings of the Royal Society, Edinburgh*, **82B**, 291–302.

Ratcliffe, P.R. (1984b) Population density and reproduction in red deer in Scottish commercial forests. *Acta Zool. Fenn.*, **172**, 191–2.

Ratcliffe, P.R.(1987) Red deer population changes in woodland and the independent assessment of population size, in *Mammal Population Studies* (ed. S. Harris). *Symp. Zool. Soc. Lond.*, **58**, 153–65.

Ratcliffe, P.R., Hall, J. and Allen, J. (1986) Computer predictions of sequential growth changes in commercial forests as an aid to wildlife management, with reference to red deer. *Scottish Forestry*, **40**, 79–83.

Red Deer Commission* (1981) *Red Deer Management: A Practical Book for the Management of Wild Red Deer in Scotland.* HMSO, Edinburgh.

Red Deer Commission (1990) Evidence to the Agricultural Select Committee, in *Land Use and Forestry*, HMSO, Edinburgh.

Red Deer Commission (1990) *Annual Report for 1989.* HMSO, Edinburgh.

*Following the introduction of the new legislation, the Red Deer Commission is now known as the Deer Commission for Scotland.

Red Deer Commission (1993) *Annual Report for 1993/94.* HMSO, Edinburgh.

Ritchie, J. (1920) *The Influence of Man on Animal Life in Scotland.* Cambridge University Press, Cambridge.

Scottish Natural Heritage (1994) Red deer and the natural heritage. Policy Paper, Scottish Natural Heritage, Edinburgh.

Staines, B.W. and Ratcliffe, P.R. (1987) Estimating the abundance of red deer and roe deer and their current status in Great Britain, in *Mammal Population Studies* (ed. S. Harris). *Symp. Zool. Soc. Lond.*, **58**, 131–52.

Staines, B.W. (1978) The dynamics and performance of a declining population of red deer *Cervus elaphus*. *Journal of Zoology, London*, **184**, 403–19.

Staines, B.W., Crisp, J.M. and Parish, T. (1982) Differences in the quality of food eaten by red deer *Cervus elaphus* stags and hinds in winter. *Journal of Applied Ecology*, **19**, 65–77.

Steven, H.M. and Carlisle, A. (1959) *The Native Pinewoods of Scotland.* Oliver and Boyd, Edinburgh and London.

Stewart, L.K. (1986) Red deer and their influence on vegetation management in northern Britain. *Deer*, **6**, 345–6.

Sykes, J.M. (1992) Caledonian pinewood regeneration: progress after 16 years of enclosure at Coille Coire Chuilc, Perthshire. *Arboricultural Journal*, **16**, 61–7.

Thompson, D.B.A., MacDonald, A.J., Marsden, J.H. and Galbraith, C.A. (1995) Upland heather moorland in Great Britain: a review of international importance, vegetation change and some objectives for nature conservation. *Biological Conservation*, **71**, 163–78.

Watson, A. and Staines, B.W. (1978) Differences in the quality of wintering areas used by male and female red deer *Cervus elaphus* in Aberdeenshire. *Journal of Zoology*, **186**, 544–50.

Watson, A. (1983) Eighteenth century deer numbers and pine regeneration near Braemar, Scotland. *Biological Conservation*, **25**, 289–305.

Watson, A. (1989) Land use, reduction of heather and natural tree regeneration on open upland. *Institute of Terrestrial Ecology Annual Report.* HMSO, London.

Welch, D., Scott, D. and Staines, B.W. (1992) Study on the effects of wintering red deer on heather moorland: report of work done April 1992–November 1992. Report to Scottish Natural Heritage, 07/91/F2A/218.

–11

Supplying primates for research

Mary-Ann Stanley and Owen Lee Griffiths

11.1 INTRODUCTION

Mauritius, a small island of 1865 km^2 in the Indian Ocean, is home to some 25 000–35 000 feral monkeys (*Macaca fascicularis*) (Sussman and Tattersall 1986). They were introduced by the Portuguese or the Dutch some time before 1606, when they were already reported in the wild. Monkeys occur in all of the remaining forested areas of the island (Figure 11.1). However, they are in their highest densities in the forests of the south-west part of the island along the escarpment of the Black River Ranges. Most of this area consists of degraded thickets and more open savannah with, at higher altitudes, areas of remaining native forest heavily degraded by Chinese guava (*Psidium cattleyanum*).

Sussman and Tattersall (1986) recorded a monkey density in prime monkey habitat of 1.30 individuals per hectare. In moister cooler upland areas over 400 m in altitude, densities are the lowest with fewer than 0.33 individuals per hectare. While monkeys prefer the lower-altitude secondary forests, they are known to migrate up to higher elevations to feed on the Chinese guavas when these are in fruit (March to June). The monkeys also feed on elements of the native fauna and flora so they are regarded as being a serious pest, not only to agriculture.

11.2 ORIGIN OF MAURITIAN MONKEYS

The origin of Mauritian monkeys is not exactly known. Sussman and Tattersall (1986), in consideration of the external morphological traits and historical situation, inferred that the monkeys came from Java. The blood protein polymorphisms of Mauritian monkeys were examined in a study conducted in 1990 (Kondo *et al.* 1991). An attempt to work out the origin of the monkeys was then made by comparing their genetic variability with

Conservation and the Use of Wildlife Resources. Edited by M. Bolton.
Published in 1997 by Chapman & Hall. ISBN 0 412 71350 0.

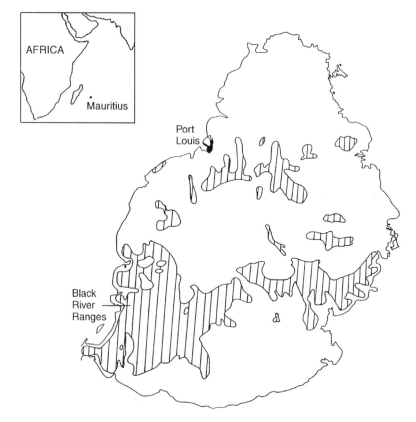

Figure 11.1 Distribution of the crab-eating macaque (*Macaca fascicularis*) on Mauritius. (Source: Bertram 1994, courtesy of the RSPCA, UK.)

those of various Asian monkey populations. The results suggested that Java was indeed the most likely origin of Mauritian monkeys. The possibility of multiple introductions from different localities was also apparent.

11.3   MONKEYS AS PESTS

Monkeys have long been regarded as serious pests in Mauritius. There are many references to this in the literature. Some writers have even suggested that the eventual Dutch abandonment of Mauritius in 1705 was due to the damage caused to plantations and food stocks by monkeys and rats (Carié 1916). Over a century ago Thompson (1880, in Sussman and Tattersall 1986) lamented the damage done by macaques to birds' eggs and seeds.

Jones (in Diamond 1987) details the significance of monkey predation on the eggs of the endangered endemic Mauritian pink pigeon, which in one area was shown to be almost 100%. That monkeys are familiar with birds' eggs was demonstrated by Jones when he placed quail eggs in cages of feral and captive-bred monkeys. The feral monkeys skilfully devoured the eggs, while captive-bred monkeys were at first frightened of the eggs and did not know what to do with them. The Mauritian Wildlife Appeal Fund, which has responsibility for restoring areas of degraded native forest, will often place scarecrows complete with a toy rifle in areas of conservation importance in an attempt to keep monkeys out of these areas.

While most references to monkeys as pests focus on their conservation significance, monkeys remain a serious pest to agriculture in Mauritius. Monkeys will attack and damage most types of crops. Most significant damage is done to sugar cane and vegetables. Young pines are also damaged. The only crop that seems to suffer no monkey damage is tea. Bertram (1994) attempted to quantify the damage done by monkeys to Mauritian agriculture. His conclusion was that monkeys cause damage costing at least £1 million annually.

11.3.1   Legal status

Understandably, monkeys have long been officially regarded as vermin in Mauritius. Monkeys were extensively hunted on the island until 1984, with sugar estates killing in excess of 1000 every year (personal communications with sugar producers in Mauritius). Monkeys remain a popular but prized food item among some ethnic groups, and monkey curry ('curry number 2' as it is locally known) remains a special treat. The decrease in the hunting of monkeys since 1984 was a direct result of the establishment of a commercial monkey farm which encouraged land owners to trap and sell live monkeys. The status of monkeys changed in Mauritius in 1995 with the passing of new legislation. From having been regarded as pests, monkeys became classified as 'proscribed' species. As a result, hunting, capture, keeping, rearing and trading in monkeys now require a permit. The only exception is that people may keep up to two monkeys as pets without a permit. The change in the legal status of monkeys is due at least in part to the fact that over the last decade monkeys have become a valuable commercial item.

11.4   MONKEYS AND BIOMEDICAL RESEARCH

For an understanding of the development of a monkey industry in Mauritius it is important to have some idea of the importance of *Macaca fascicularis* to the world's biomedical research industry. This species is the preferred research monkey with some 10 000–12 000 being used annually. Of this number 70% are captive bred. Most of the captive breeding occurs in

Mauritius, Indonesia, Philippines, China and Vietnam. The use of wild-caught monkeys is now restricted in Europe, and the only substantial market for wild-caught monkeys is the USA.

Mauritian monkeys are known for their very good health profile and are free of herpes B virus, SIv, SRV1, SRV2, STLV1, filovirus, rabies, measles and simian malaria. This means that Mauritian monkeys for many researchers are the monkey of choice. The fact that they are not indigenous to Mauritius is an added reason for the biomedical research industry to prefer Mauritian monkeys. They are used for vaccine work including the testing of polio vaccines, AIDS work and a host of other biomedical research projects.

11.4.1 Commercial developments

The monkey farming industry in Mauritius was established in 1984 with the granting by the Government of Mauritius of an export licence and quota to Bioculture (Mauritius) Ltd. Since that time the industry has grown to the stage where there are now two companies exporting monkeys and six additional companies acting as contract breeders for the exporting companies. Total direct employment in the monkey industry in Mauritius is about 220. At any given time there are over 8000 monkeys in captivity, including over 4000 breeding females and breeding males with the balance being captive-bred offspring.

The protocols for dealing with monkeys are virtually the same for all the companies involved and are based on International Primatological Society Guidelines. The overall management is governed by the Veterinary Services and the National Parks and Conservation Service, which is the CITES management authority for Mauritius.

11.5 TRAPPING MONKEYS

Monkeys in Mauritius are trapped in those areas where they represent a threat to agriculture or to native fauna and flora. Two types of traps are used. Type 1 traps are designed for catching monkeys individually and have a door rigged to close when a monkey enters the trap and pulls on the bait. Type 2 traps are gang traps which are large (6 m × 6 m × 3 m) and manually operated. A trained trapper is positioned in a hide 20 m from the trap. These traps are baited with sugar cane, and groups of animals are caught when the trapper pulls the door shut. The trapped monkeys are transferred to field transport cages.

All monkeys caught are taken by jeep to reception facilities, usually no more than 30 minutes away from the trapping sites. There they are released into large holding cages with animals of similar size being grouped to await their first veterinary inspection. The close proximity of trapping and reception areas is a great advantage in Mauritius. In south-east Asia most trapping areas are many days' travel time from the reception facilities, and monkeys consequently suffer great travel stress. They are also exposed to disease from the many humans with whom they come into contact between capture and arrival at the end facility – which is usually in a capital city. There are no such problems in Mauritius.

At the reception facility, newly arrived monkeys are inspected and given a range of vaccines and treatments for endo- and ectoparasites. They are also given a tattoo number to allow for a detailed medical record to be kept on each animal. Animals then undergo a 3-month quarantine period during which time they are given five TB tests. Following this, they are selected either for breeding or to be exported. Animals destined for breeding are also given a chest X-ray.

11.6 BREEDING OF MONKEYS

Monkeys are bred in harem groups consisting of 45 females and three males. They are placed in two cages, each 6.5 m × 6.5 m × 2 m (height) with a connecting corridor. The cages are fitted out with sleeping platforms and a variety of environmental enrichment devices including tunnels, trampolines and swings. Water is supplied ad lib by watering nipples. Animals are each fed 150 g of pelletized primate diet and 250 g of mixed fruits and vegetables daily. Breeding success ranges from 75 to 88 saleable offspring at 18 months from 100 females. Newly formed groups are made up of feral animals of various but unknown ages. Breeding success peaks within the first few years, and gradually declines over time as the colony ages. There are now breeding colonies in Mauritius that are 10 years old.

The gestation period is 160 days. Births are recorded daily and babies are tattooed at 2 weeks of age. Weaning takes place at 7–8 months and at a minimum weight of 1.2 kg. This is about the natural weaning age in the wild.

Weaned, captive-bred offspring are then placed in growing cages which contain groups of 40 animals of mixed sex. They are fed the same amount of food as the breeders except that primate pellets containing 25% protein are used. Breeders' pellets contain 18% protein. Males mature at 3.5–4 years of age and at approximately 4 kg in weight. Females mature at 3 years and at a weight of 2.8–3 kg.

Captive-bred animals are usually exported at 18–24 months, though some animals are reared to sexual maturity as this is required by some customers. Mauritius exports monkeys mostly to the following countries: USA, Canada, UK, France, Belgium, Italy and Israel.

The industry is now starting to set aside captive-bred offspring in order to produce F2 generation animals. This is essentially in response to UK recommendations that breeders move towards 'closed cycle breeding'.

11.7 MONKEYS AND CONSERVATION

From its humble beginnings in 1984, the monkey industry has grown to become very important to Mauritius. Monkeys are now the biggest live-stock industry in Mauritius in terms of export earnings (over Rs100 million per annum). As mentioned above it is also an important employer providing over 220 direct jobs and much more indirect employment.

The industry also provides the largest source of locally derived conservation funding in Mauritius. For each monkey exported (captive-bred or feral) the industry pays $US50 to the Ministry of Agriculture's Conservation Fund. As Mauritius now exports over 4000 monkeys a year, this fund receives in excess of $US200 000 per year. The funds have been used to top up a World Bank loan to set up Mauritius' first National Park. In addition these funds are the only funds available for the ongoing costs of running the National Park. Monkey funds are also being used to fund the restoration of large areas of degraded native forest by paying for weeding and the erection of fences to exclude introduced deer and pigs.

Ironically, monkeys, which have for so long contributed to the destruction of the native fauna and flora of Mauritius, are now helping to save what is left. Equally ironic, the existence of a monkey industry has provided some *de facto* protection for monkeys because all the large sugar estates now actively discourage monkey shooting, preferring to trap monkeys for sale. It must be clarified here that monkey shooting (until the passing of the 1995 legislation) was really sport hunting that doubled as pest control (albeit with limited success). If the lucrative market for feral monkeys were to disappear, there would, without doubt, be a resumption of sport/pest control shooting of monkeys.

11.8 PROBLEMS AND PROSPECTS

The monkey industry in Mauritius has been able to grow to its current size because it has had the full support of successive Mauritian governments. They have recognized its importance to the economy and to conservation. Also they have recognized that the removal of pest monkeys for export or breeding is a sensible way of dealing with the problem of feral monkeys in Mauritius.

In the early 1990s the RSPCA (UK) offered to advise the Mauritian Government on ways of dealing with the monkey problem that might be more humane than allowing the animals to be the basis of an export industry for biomedical research. In 1994 the Government of Mauritius accepted the RSPCA's offer and the RSPCA in turn engaged a consultant from the Zoological Society of London to examine various control methods. Complete elimination of the population was found not to be feasible. It was also concluded that the preferred option of controlling reproduction by contra-

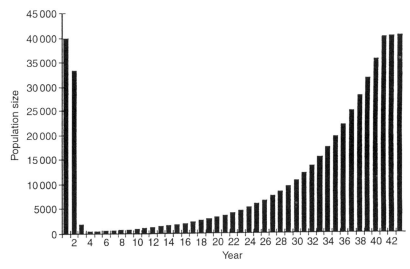

Figure 11.2 Reproductive potential of the crab-eating macaque. Population model-ling indicates that, in a population with a carrying capacity of 40 000 and density-dependent breeding, a population of 40 000 macaques will return to carrying capa-city 38 years following a 99% reduction in numbers. (Source: Bertram 1994, courtesy of the RSPCA, UK.)

ception was also unlikely to be effective as there was no realistic possibility of applying contraception by any method to all female monkeys. Population modelling indicated that, in a population with a carrying capacity of 40 000 monkeys and density-dependent breeding, a population of 40 000 monkeys would return to carrying capacity 38 years following a 99% reduction in numbers (Figure 11.2). The conclusion was that no suitable control measure exists at this stage (Bertram 1994).

If any threat exists to the long-term viability of a monkey farming industry in Mauritius (or indeed anywhere outside of Europe or the USA), it is from activist groups who are opposed to the trade. It is to counter just such a threat to the supply of monkeys to Europe that the EU has set up a commission to look into the breeding of monkeys in Europe. The commis-sion would like to see Europe self-sufficient in monkeys. For the same reason there has been an increase in captive breeding in the USA. However, neither the USA nor Europe can compete commercially with suppliers based in tropical Third World countries where there is no need for costly indoor facilities and where labour is much cheaper. It is an irony, though, that European animal rights' groups may cause a total transfer of monkey breeding from the tropics to the USA and Europe with a consequential loss of vital conservation funding in places like Mauritius and an undoubted return to shooting of pest monkeys in Mauritius.

It is appropriate to conclude this chapter by quoting Kavanagh (1983):

> South East Asia's long-tailed macaques have been established far to the west of their natural range on Mauritius. They pose a serious threat to some of the rare birds of Mauritius by eating their eggs and nestlings. If monkeys have to be taken for research, these would seem to be the ideal candidates.

REFERENCES

Bertram, B. (1994) *Monkeys in Mauritius: Potential for Humane Control*. Report commissioned by the RSPCA for the Government of Mauritius.

Carié, P. (1916) L'acclimitation à L'Ile Maurice. *Bull. Soc. Natl Acclim.*, Paris, 63. (In Diamond 1987.)

Diamond, A.W. (1987) *Studies of Mascarene Island Birds*. Cambridge University Press, Cambridge.

Kavanagh, M. (1983) *A Complete Guide to Monkeys, Apes and other Primates*. Oregon Press, London.

Kondo, M., Kawamoto, Y., Noza, K., Matsubayashi, K., Watanabe, T., Griffiths, O. and Stanley, M. (1991) Population genetics of crab-eating macaques (*Macaca fascicularis*) on the island of Mauritius. *American Journal of Primatology*, 29, 167–82.

Sussman, W. and Tattersall, I. (1986) Distribution, abundance and putative ecological strategy of *Macaca fascicularis* on the island of Mauritius, Southwestern Indian Ocean. *Folia Primatol.*, 46, 28–43.

12

Some veterinary risks and public health issues in wildlife utilization

Michael H. Woodford

12.1 INTRODUCTION

Human beings, wild animals and the domestic livestock derived from them have shared the same environments throughout most of their evolution. In consequence, many parasitic and pathogenic organisms have developed life cycles which take advantage of this physical proximity. The diseases and infections which are naturally transmitted between humans and animals are called **zoonoses**.

Hardouin (1995) refers to the many small animals eaten by subsistence hunters as 'minilivestock' and points out that relatively little is known about the health aspects of this range of animals – either in regard to animal health or the associated risks to human consumers. In this chapter, the emphasis is on the consumptive use of large mammals but the significance of non-food-borne risks is also considered.

12.2 NON-FOOD-BORNE RISKS

12.2.1 Culture and way of life

The opportunity for acquiring a zoonotic infection may be greatly influenced by cultural and lifestyle factors, and special risks are encountered by those living in crowded and insanitary conditions. Pregnancy, adolescence, old age and immunosuppression also increase the disease risk situation (Morosetti and Mole 1992).

During normal working and recreational activities in the field, which may include farming, forestry, hunting and fishing, humans may come into contact with the arthropod vectors of zoonotic diseases such as visceral

Conservation and the Use of Wildlife Resources. Edited by M. Bolton.
Published in 1997 by Chapman & Hall. ISBN 0 412 71350 0.

leishmaniasis (Kala azar), Latin American trypanosomiasis (Chaga's disease), borreliosis (Lyme disease), sleeping sickness (African trypanosomiasis) and many arboviral diseases such as yellow fever and Kyasanur Forest disease. Humid, low-lying lake margins and river banks provide the habitat for wild animal disease reservoirs and the arthropod vectors. Rice-field workers in these areas are frequently infected with leptospirosis (through contact with rodent urine) and tularaemia (through handling and skinning small mammals). Dryland and forest biotopes harbour different pathogens, different wild disease hosts and different vectors which can transmit sylvatic plague, histoplasmosis, rickettsiosis, viral haemorrhagic fevers and rabies, all of which present a potential hazard to those humans who work and travel in these environments.

12.2.2   Conservation and recreation

Translocation is a popular conservation technique for the reintroduction or restoration of depleted wildlife populations. However, the vast majority of translocations are made solely for sporting purposes (Griffith *et al.* 1989) and seldom is it appreciated that each translocated animal is not just a representative of a single species but is rather a biological package containing an assortment of viruses, bacteria, protozoa, helminths and arthropods, many of which have the potential to cause or transmit disease (Nettles 1988; Woodford and Rossiter 1994). The habituation of the great apes (*Gorilla gorilla*, *Pan troglodytes* and *Pongo pygmaeus*), so that tourists can approach the animals to within a few metres in their natural habitat, must present a grave disease risk and one that is especially significant in the case of those taxa which are classified as 'endangered'. The great apes are susceptible to human diseases such as influenza, measles, mumps and tuberculosis, while the humans themselves run a rather smaller risk of acquiring hookworms (*Ancylostoma* sp., *Necator* sp.), salmonellosis, shigellosis and infectious hepatitis from the apes (Fowler 1986).

12.2.3   National parks and protected areas

Animals confined in national parks often experience disease problems, some of which can be zoonotic, when their population density increases and climatic and social stresses supervene. Increasing contact and transmission of disease by trespassing domestic animals, mainly cattle and semi-feral dogs, has resulted in several recent disease outbreaks, for example, bovine tuberculosis in Cape buffalo (*Syncerus caffer*) in Kruger National Park, South Africa (Bengis, pers. comm.) and canine distemper in lions (*Felis leo*) in the Serengeti National Park, Tanzania (Munson *et al.* 1995). In both cases the disease has become enzootic and control will prove difficult.

12.2.4   International trade

The international trade in wild animals for exhibition in zoos, for the pet trade and for medical research is subject to strict regulations in most countries. But even these break down on occasion, as evidenced by the recent occurrence of Ebola virus infection in macaques (*Macaca fascicularis*), imported into USA from the Philippines for medical research purposes (Morosetti and De Nardo 1990; WHO 1990) and the outbreak of African horse sickness (AHS) in Spain which followed the import of two zebra (*Equus burchelli*) from Namibia for release in a Wildlife Park (OIE 1987). The imposition of rigorous quarantine regulations has reduced but not eliminated the disease risks which accompany imported wild animals.

12.3   FOOD-BORNE RISKS

12.3.1   Introduction

Systems of wildlife exploitation for meat production vary widely from large-scale harvesting of wild animals in order to supply a luxury, urban market or for export; to the so-called subsistence hunting for bush meat, practised in many parts of rural Africa, Asia and South America, where product hygiene is likely to be non-existent.

The former usually depends upon the availability of a croppable species in sufficient numbers on suitable terrain and near enough to a centre of human population whose income permits the purchase of fresh game meat at realistic prices. In this instance, the wildlife meat has to compete in price with the butchered meat of domestic animals and must achieve comparable standards of hygiene. When such game meat is destined for export, processing procedures and standards of hygiene are set by the importing country. These may vary somewhat depending on the country concerned but are usually very strict.

Between these two extremes there have been in Africa both private and government-sponsored cropping schemes. These have variously involved the harvesting of zebra in Tanzania, impala (*Aepyceros melampus*) in Zimbabwe, Cape buffalo in Mozambique (mainly to feed the army but see Plate 7) and elephant (*Loxodonta africana*), and several species of antelope in South Africa (for export). Standards of hygiene in the field vary, but in most cases it is in the interest of the cropping agency to maintain reasonable standards so as to extend the shelf-life of the product and retain the goodwill of its customers. In some countries, for example Zimbabwe and South Africa, where game cropping operations are on a considerable commercial scale, the Hygiene Division of the Department of Veterinary Services has become increasingly involved in the control of slaughter, dressing, refrigeration and inspection, and of the management of game-processing houses. This official involvement is designed to prevent the spread (by meat products) of food-borne infections to the human consumers and of diseases communicable to domestic animals.

Meat from these cropping operations is usually sold locally, and with the exception of certain constraints applied for the control of foot-and-mouth disease (FMD) in specific areas, there are usually few controls placed on within-country movements of the products (meat and skins).

Only in the case of relatively large-scale commercial cropping operations for a luxury urban market has an evaluation of the problems of hygienic wildlife meat production been made.

12.3.2 Commercial cropping for meat

One of the more recent examples has been the pilot harvesting exercise carried out on the Anti-Kapiti Plains, near Nairobi in Kenya, by the FAO/ UNDP Kenya Wildlife Management Project (KWMP) during the mid-1970s. Much of what follows derives from experience gained on that project but the hygiene problems encountered then are likely to be common to most systems of commercial wildlife meat harvesting, especially in the tropics.

12.3.3 Ante-mortem inspection

Contrary to what one might expect, the marksman engaged in cropping wild animals in the field has a much better opportunity to carry out an ante-mortem inspection than does the veterinarian whose duty it is to inspect a crowded bunch of cattle in a lairage at a commercial slaughter-house. This is because the cropper has a chance to study the uninhibited behaviour and demeanour of the target animals before he selects one and fires his shot. He can reject any animal that shows obvious signs of disease which could present a hygiene hazard in the field abattoir. In order to do this he will require training in the recognition of the overt signs of disease conditions which, in addition to resulting in the condemnation of the carcase and waste of the abattoir staff's time, may contaminate the abattoir and equipment with pathogenic organisms.

12.3.4 Slaughter

Experience has shown that bruising due to inaccurate shooting can be a major cause of carcase condemnation. Apart from the obvious welfare concerns, therefore, the correct choice of rifle calibre and bullet weight is important when cropping large mammals for meat. Shotguns, although often used for night shooting of small antelopes, are generally unsuitable because the pellet pattern is uneven and penetration of the abdominal cavity is frequent. The subject of satisfactory slaughter techniques has been explored in detail in the report of the FAO/UNDP KWMP Veterinary Section (1976).

12.3.5   Skinning

It is undesirable to wet the hair of a carcase to lay the dust before skinning because this practice tends to increase surface contamination. Hygienic skinning is thus very difficult to achieve when the animals have been shot when wet, and cropping operations should be suspended during or after rain.

12.3.6   Evisceration

Much controversy exists as to what should be regarded as an allowable time-lapse between shooting and evisceration. Commercial domestic slaughterhouse managers usually insist on a maximum evisceration time of 45 minutes for cattle and 60 minutes for sheep and goats. Tests carried out in Kenya by KWMP, mainly on wildebeest (*Connochaetes taurinus*) carcases, have shown that an evisceration time of up to 90 minutes is acceptable under field conditions for animals of that size and that such a time-lapse does not result in unacceptable contamination with enterobacteria. It is, however, very important that the animals are properly bled, that carcase contamination is minimal during evisceration and that the carcases are subjected to rapid cooling.

12.3.7   Post-mortem meat inspection

Inspection procedures based on conventional domestic meat inspection are satisfactory for the inspection of wildlife meat in field abattoirs (Plates 5 and 6). Meat inspectors trained in domestic meat inspection routines need very little extra training to convert them into wildlife meat inspectors. The major hygiene problems for African antelope meat destined for the urban luxury market or for export are muscular cysticercosis and bruising due to inaccurate shooting. In the Kenya operation it was found that all the parasitic cysts seen in the muscles of wild antelopes shot in the project area were the larval stages of specific tapeworms of lions and hyaenas (*Crocuta crocuta*), which are believed to be non-pathogenic for humans. In no case were they found to be the cysts of *Cysticercus bovis*, the intermediate stage of the human beef tapeworm *Taenia saginata*. Accordingly, only those carcases found to be infested by more than five cysts (including cardiac cysts) on conventional meat inspection cuts were rejected on aesthetic grounds. Carcases infested by five or more cysts were frozen in a commercial deep freeze for 14 days at $-20°C$ and subsequently sold as manufacturing meat, but this procedure was found to be barely economic. It is however, now believed that conventional domestic animal meat inspection cuts cannot be relied upon to reveal the extent of parasitic cysticercosis in wild antelope meat.

Bruising, due to inaccurate shooting, was responsible for the condemnation of about 30% of the inspected wildebeest carcases but although this meat was unsuitable for sale in a luxury or export market, it was perfectly safe and acceptable for immediate consumption by local people in the cropping area.

Conventional meat inspection, modified to suit the circumstances, is both necessary and possible for large scale commercial cropping operations. It will continue to be impossible in the case of the subsistence hunter until his activities are restrained. Education of the consumer, too, will play a part but this will take time, as evidenced by the apparent abandon with which pastoral people, who well recognize the seriousness of the disease, will consume meat from an anthrax carcase.

12.4 PRESERVATION OF WILDLIFE MEAT

Adequate cooling of wildlife meat in the field to 7°C as soon as possible, followed by refrigeration at 4°C, is essential if keeping quality is to be maintained. However, if cropping operations take place in areas so remote from town markets or refrigeration facilities that the production of fresh meat for early sale is impractical, methods of meat preservation other than chilling and freezing become necessary. Alternative methods of preserving meat in the field include smoking, salting and drying or a combination of all three. These techniques are well understood by traditional subsistence hunters and are also used by trophy hunters to preserve skins and other trophies. It must, however, be remembered that probably one of the most important and dangerous zoonoses that is likely to be transmitted by fresh wildlife meat is trichinosis and that smoking, salting or drying cannot be relied upon to sterilize meat infested with this parasite. *Trichinella* sp. is likely only to infest the meat of wild pigs, rodents and carnivores, animals which, though unlikely to be cropped commercially, may form important components of bush meat harvested by subsistence hunters.

12.5 SAFE DISPOSAL OF CONDEMNED MEAT AND OFFAL

Under field conditions the disposal of condemned meat and offal presents special problems. Often the subsoil is rocky, making burial at a sufficient depth impractical, and seldom is there adequate firewood available to burn the waste material. Scavengers, avian, mammalian and human, are quick to take advantage of the situation and even material not normally considered to be edible will disappear without trace. While it is probable that passage through the guts of vultures sterilizes meat as far as metazoan parasites are concerned, this is less likely when the pathogen is a long-lived bacterial spore like that of *Bacillus anthracis* (anthrax) (Houston and Cooper 1975). Luckily the presence of a case of anthrax at post-mortem in a cropped mammal is very unlikely.

Meat and offal condemned on account of generalized muscular cysticercosis (muscle cysts) or hydatid cyst infestation are probably sterilized by passage through vultures, provided such material is put out in the bush early enough in the day for vultures to find and consume before dark. A further provision is that the supply of condemned material is not so great that the avian scavengers become sated and are unable to eat it all. Jackals (*Canis* sp.) and feral domestic dogs may find the condemned material and compete successfully with the vultures for it. In this case there is a grave danger of further contaminating the local canid population, both wild and domestic, with *Echinococcus* sp., the intermediate cystic stage of which often infects humans, with unpleasant consequences. Lions and hyaenas are more likely to visit the condemned meat dump site at night so they, too, can acquire heavy infestations of tapeworms by eating this concentrated, infected material. Human scavengers are at less risk from parasites because the larval tapeworm cysts in wildlife meat are unlikely to be those of human tapeworms, and the stolen meat will probably be cooked, but the unhygienic conditions of a condemned meat dump are hardly conducive to health and the humans are often accompanied by domestic dogs.

If conditions are suitable for burial of condemned meat and offal, it must be remembered that hydatid cysts in livers and lungs can remain viable for up to nine days underground and that hyaenas are capable of digging to considerable depths. Hydatid-infested livers and lungs left on the surface of the ground at normal temperatures cannot be considered inactive until at least a week has elapsed, although invasion by fly larvae will hasten the process of destruction.

A method of sterilizing hydatid cysts in offal which has been found to be effective in Uruguay, where burning or burial is impractical, is to cut the cysts open and immerse them and the organ in which they occur in saturated salt solution. This procedure would be practical in a field cropping operation, or indeed at a village slaughterhouse, since all that is required is a 200-litre (44-gal) petrol drum, water and a supply of salt.

12.6   TROPHY HUNTING, SUBSISTENCE HUNTING AND BUSH MEAT

The exploitation of the wildlife resource by subsistence hunters is often on a commercial basis, too. In the forested areas of West Africa, south-east Asia and northern South and Central America, the rural people are generally not hungry and while they may hunt to provide their families with much needed protein, they also do so for profit. The animals they kill comprise a wide variety of taxa which may range from large mammals to small rodents, birds, reptiles, amphibia, fish and invertebrates. Much of this material is temporarily preserved by smoking and drying over open fires and is transported to the towns and villages where it commands a high price. Hygiene is generally non-existent, and veterinary issues receive little consideration. As remarked above, while smoking and drying will extend the shelf life of bush meat, it cannot be relied upon to sterilize it of parasites or other pathogens. In this context, *Trichinella* sp. has been mentioned as a dangerous parasite which has infected people who have eaten undercooked wild pig meat, especially in East Africa (Nelson 1983). Wild pig meat (*Sus scrofa*) can also be infested by the cysts of the human tapeworm (*Taenia solium*), especially in south-east Asia. Much polar wildlife is infested by *Trichinella* sp., including polar bears (*Thalarctos maritimus*), seals (*Phoca greenlandica*) and walruses (*Odobenus rosmarus*), and people can become infected by eating raw or undercooked meat of these animals. Trophy hunters who eat the flesh of bears, wild pigs and carnivores are at risk from trichinosis, and African wild pig trophies imported into Europe have been implicated in the introduction of African swine fever virus (ASF) (Mantovani and Leporati 1976).

Toxoplasma gondii is a very widespread protozoal parasite which has been reported to infest at least 200 species of mammal worldwide (Morosetti and Mole 1992). Outbreaks of toxoplasmosis have followed the ingestion of undercooked deer meat (Sacks *et al.* 1983) and kangaroo meat (*Sydney Morning Herald*, 25 October 1995). Those who hunt, for subsistence or sport, are at risk from infection by tularaemia when they skin and process small mammals; similarly they may infect themselves with hydatid cysts if they swallow the eggs of *Echinococcus* sp. (a very small tapeworm) while skinning carnivores. Trophy hunters who skin and eat their quarry are, of course, exposed to the same zoonotic diseases and parasites as subsistence hunters, especially if they fail to take elementary hygiene precautions.

Very little is known of the incidence and prevalence of zoonotic diseases in subsistence hunters. This is surprising when one considers that as many as 80 million wild grasscutters (*Thryonomys* sp.), representing 300 000 metric tons of meat, are killed and eaten annually in West Africa (Jori *et al.* 1995) and the estimated annual consumption of bush meat by the people of Gabon is 17 million kg (Steel 1994).

12.7   INTERNATIONAL REGULATIONS

12.7.1   European Union

Of the 52 African countries, only Botswana, Namibia, Swaziland, Zimbabwe and South Africa are permitted to export the meat of wild, cloven-hoofed animals to the member countries of the European Union (EU). In these five cases, the meat of wild pigs is excluded (ASF risk) and all meat must be deboned and must exclude offal. Twenty-two other countries of the world can export the meat of wild, cloven-hoofed animals to the EU but the export of wild pig meat is not allowed from China, Chile, Greenland and the former USSR (EU Council Directive 1992).

The requirements of most EU member-country veterinary and public health departments specify that game meat carcases must be certified to have been individually inspected by an official veterinarian, and to have been dressed, deboned and packaged in a game meat plant approved by the importing country's veterinary authorities. None of the meat should have been derived from a specified 'FMD risk' area. In practice, in Africa this usually means an area in which African Cape buffaloes are found, since these animals are known often to be the symptomless carriers of FMD virus.

The EU requires that the fresh meat of wild cloven-hoofed animals which are harvested from fenced game ranches (and therefore considered as 'farmed') shall be subjected to the same obligatory sampling protocols as domestic stock for environmental contaminants, hormones, antibiotics and pesticides.

At present the EU has no legislation covering the import of fresh meat from exotic species such as crocodiles, snakes or elephants, and jurisdiction for such imports rests solely with the veterinary and public health authorities in each member state.

Outside the EU the conditions of import of wild animal products are prescribed by the importing country. These may vary from no conditions at all – not even veterinary or public health certification – to strict prescriptive conditions which are equivalent to those required for domestic meat hygiene and are often modelled on the EU Directives.

12.7.2 USA and Canada

In the USA the situation with regard to game animal products is rather different. Federal regulations, designed to control poaching rather than public health, forbid the sale of edible parts of resident wild animals or migratory game birds (waterfowl, rails, coots and doves). These resources can, however, be consumed in the home of the hunter or given away. Non-edible parts – horns, skins and trophies, with the exception of plumes for millinery purposes – can be sold, but this trade is subject to state legislation. The USA imposes very strict regulations on the import of wild game meat from any source, and so stringent are the required laboratory tests that most exporters seek a less demanding market.

In Canada it is illegal to sell the meat or products of any wild animal or bird. However, some native Canadian groups are permitted to sell wild game meat. 'Farmed' venison (deer meat) can be sold at 'farm gate' sites without any veterinary or hygienic inspection. In many rural communities in Canada, 'farmed' venison can be sold after slaughter at an abattoir which has been inspected by the provincial health inspector. No federal inspection is required. In the larger Canadian cities there are by-laws requiring inspection of 'farmed' venison, by a Federal Inspector, after slaughter in a registered slaughterhouse. Game meat (usually 'farmed' venison) destined for transport across provincial boundaries or international borders similarly has to be slaughtered at one of the few federally registered plants. The veterinary and hygienic standards at such plants are said to be strict and would probably meet EU requirements. Game birds may not be sold if they are wild-shot but can be traded if they are farm-raised for commercial purposes. There are no inspection requirements for these.

12.7.3   New Zealand and Australia

New Zealand has declared 'farmed' deer to be domestic animals and these must now be slaughtered in Deer Slaughter Premises (DSPs) which comply with the standards of export abattoirs for domestic animals and also with those of the EU for a 'wild game processing house'. These standards are governed by the New Zealand legislation and regulations relating to meat processing and marketing and by the import protocols of importing countries. The reason why deer must be processed in DSPs is that New Zealand venison is still classified as 'game meat' by the EU rather than 'farmed meat' and as a result can enter the EU without duty or quantitative restrictions. The export of velvet (dried developing antlers) is subject to export protocol standards and to health regulations relating to the processing of a food product. These are primarily hygiene-based. The majority of the velvet is exported to South Korea where it is used in traditional human medicine.

Australia, like New Zealand, has the great advantage that it is free from many of the important infectious diseases of wild and domestic animals, e.g. FMD, ASF, AHS and trichinosis. Australia does, however, have bovine tuberculosis in feral water buffaloes and feral pigs but this disease has been almost eliminated from the water buffalo and in the wild pig is considered to be a 'dead-end' disease.

Feral water buffaloes in Australia are classed as slaughter animals and if they are to be used for human consumption they must be transported live to an abattoir. Feral pigs and goats are classed as game animals, as are kangaroos. In Australian legislation a game animal is defined as a wild animal, other than a bird or rabbit, that has been killed in its habitat by a shot from a firearm. At game processing establishments officers of the Australian Quarantine and Inspection Service carry out post-mortem inspections and issue the necessary game-meat export documents. In recent years Japan has been the most regular buyer of kangaroo game meat but a wider Asian market uses kangaroo meat as pet food (Ramsey 1994). The bulk of kangaroo meat is sold on the domestic market as pet food but sale for human consumption is now legal in all Australian states and territories (Dee 1995).

Kangaroos suffer from few diseases normally associated with domestic animals and present little danger to human health. They do, however, carry a number of parasitic worms, two of which are of aesthetic importance (Andrew 1988; Hopwood 1988). *Dirofilaria roemeri* is found in the sub-cutaneous tissues surrounding the stifle joint of the hind leg and although the parasite has no public health significance, its presence has caused some export problems. The other parasite is *Progamotaenia festiva*, a tapeworm which commonly infests the bile ducts in the liver. This parasite has no public health importance since kangaroo livers are normally not saved for human consumption.

12.8   CONCLUSIONS

Throughout the world, the main legal constraints on the export and import of meat derived from wild ruminants are the regulations controlling FMD, tuberculosis and brucellosis. In the case of pigs, both wild and domestic, the diseases are ASF, classical swine fever (CSF), FMD, brucellosis and trichinosis, and for equids, AHS. Several of these infections, such as FMD, ASF and AHS, often have very little effect on their wild hosts but can cause serious and sometimes fatal disease when transmitted to domestic stock. It is assumed that during the process of domestication, domestic animals have lost their innate resistance to some of the pathogens carried with impunity by their wild ancestors. Conversely, humans are sometimes at greater risk from parasitic infections derived from domestic stock than they are from those carried by wild animals. The phenomenon can probably be ascribed to humankind's 10 000 years of close association with domestic livestock, but while the risks of zoonotic infections derived from domestic livestock are generally well known, those of wild animals, especially of the more unconventional meat sources, are much less familiar and more research is needed into their prevalence and incidence.

REFERENCES

Andrew, A.E. (1988) *Kangaroo Meat: Public Health Aspects*. Australian Quarantine and Inspection Service, Canberra.

Dee, C. (1995) Industry responsibility and expectations in processing and marketing kangaroo meat for human consumption, in *Conservation through Sustainable Use of Wildlife* (eds G. Grigg, P. Hale and D. Lunney). Centre for Conservation Biology, University of Queensland, Brisbane, pp. 230–2.

EU Council Directive 92/45/EEC, 16 June 1992, on public health problems relating to the killing of wild game and the placing on the market of wild game meat.

FAO/UNDP Kenya Wildlife Management Project KEN/71/526. Report of the Veterinary Section, Project Working Document No. 10, Sept. 1976. FAO, Rome.

Fowler, M.E. (ed.) (1986) *Zoo and Wild Animal Medicine*, 2nd edn. Morris Animal Foundation, Philadelphia.

Griffith, B., Scott, J.M., Carpenter, J.W. and Read, C. (1989) Translocation as a species conservation tool: status and strategy, *Science*, **245**, 477–80.

Hardouin, J. (1995) Minilivestock: from gathering to controlled production. *Biodiversity and Conservation*, **4**, 220–32.

Hopwood. P.R. (1988) Kangaroos as game meat animals: carcass meat yields and meat inspection. *Australian Zoologist*, **24**, 169–76.

Houston, D.C. and Cooper, J.E. (1975) The digestive tract of the white-backed griffon vulture and its role in disease transmission among wild ungulates. *J. Wildl. Dis.*, **11**, 306–13.

Jori, F., Mensah, G.A. and Adjanohoun, E. (1995) Grasscutter production: an example of rational exploitation of wildlife. *Biodiv. Conserv.*, **4**, 257–66.

Mantovani, A. and Leporati, L. (1976) On the essential role of intergovernmental agencies in controlling animal diseases in *Wildlife Diseases* (ed. L.A. Page). Plenum Press, New York, pp. 385–92.

Morosetti. G. and De Nardo, P. (1990) Aggiornamenti su casi di infezione da virus di Ebola. *Il nuovo Progresso Veterinario*, **11**, 415–20.

Morosetti, G. and Mole, S. (1992) *Notes on the Role of Wildlife in the Epidemiology of Zoonoses.* ISS/WHO/FAO-CC/IZSTe/92.19, Superiore di Sanita, Laboratorio di Parassitologia, Rome.

Munson, L., Appel, A.P.G., Carpenter, M.A., O'Brien, S.J. and Roelke-Parker, M. (1995) Canine distemper in wild felids, in *Proc. Joint Conf. Amer. Assoc. Zoo Vets and Amer. Assoc. Wildl. Vets.* East Lansing, Michigan, pp. 135–6.

Nelson, G.S. (1983) Wild animals as reservoirs of parasitic diseases in Kenya, in *Proc. 10th Meet. World Assoc. Advanc. Vet. Paras. and Paras. Zoonoses* (ed. J.D. Dunsmore), pp. 59–72.

Nettles, V.F. (1988) Wildlife relocation: disease implications and regulations. *Proc. 37th Ann. Conf. Wildl. Dis. Assoc.*, Athens, Georgia, p. 52.

OIE (1987) Epizootiological Information, Nos ESP/87/1/117, ESP/87/2/119, ESP/87/3/121, ESP/87/4/127, ESP/87/5/142, and ESP/87/6/145, Office Internationale des Epizooties, Paris.

Ramsey, B.J. (1994) *Commercial Use of Wild Animals in Australia.* Bureau of Resource Sciences, Australian Government Publishing Service, Canberra.

Sacks, J.J., Delgado, D.G., Cobel, H.A. and Parker, R.L. (1983) Toxoplasmosis infection associated with under-cooked venison. *Amer. J. Epidemiol.*, **118**, 832–8.

Steel, A.E. (1994) A study of the value of bush meat commerce in Gabon. Programme Report, WWF, Libreville, Gabon.

WHO (1990) *Weekly Epidemiological Record*, **24**, 185.

Woodford M.H. and Rossiter P.B. (1994) Disease risks associated with wildlife translocation projects, in *Creative Conservation: Interactive Management of Wild and Captive Animals* (eds P.J.S. Olney, G. Mace and A.T.C. Feistner). Chapman & Hall, London, pp. 178–200.

13

Conservation and captivity

13.1 INTRODUCTION

Non-domestic animals are held captive under a very wide range of circumstances, and the distinction between captivity and freedom can become blurred in protected areas, where animals may be provisioned to some extent, and where human activities prevent them from going where they would like to go. At this end of the continuum, captivity may not be decided by bars or fences but by other means of restraint and control. A more clear-cut distinction can be made between animals which are kept within their natural habitats (*in situ*) and those which are kept off site or *ex situ*.

Ordinarily, wildlife conservation means trying to keep wild animal populations in their natural habitats so the conservation value of keeping animals *ex situ* is assessed in terms of its contribution (realized or potential) to that end. This chapter will examine *ex situ* captivity and its relevance to conservation.

13.2 THE CONSERVATION BREEDING SPECIALIST GROUP (CBSG)

The Species Survival Commission of IUCN has many specialist groups. Established in 1978, the CBSG is the major organization involved in global coordination of *ex situ* conservation. It is funded by voluntary donations, and forms the link between the global zoo network and the IUCN conservation network. There are currently more than 600 professionals from 150 countries working as volunteer members of this non-profit foundation (CBSG 1995).

The main activities of the CBSG are shown in Box 13.1.

Conservation and the Use of Wildlife Resources. Edited by M. Bolton.
Published in 1997 by Chapman & Hall. ISBN 0 412 71350 0.

Box 13.1   The functions and processes of the Conservation breeding Specialist Group (CBSG)

Communication
The CBSG serves as an information clearing house and publishes a quarterly newsletter, the *CBSG News*, which is circulated to some 7000 researchers and others world-wide

Training
Provision of general and specific training in addition to conducting workshops organized around the processes listed below

Population and Habitat Viability Assessments (PHVAs)
Take the form of workshops held within a range country of the species in question. Available information on the species, its habitat and requirements are fed into a computer program for analysis. The program used so far is a stochastic population simulation model, VORTEX, which was specially developed for the purpose

Conservation Assessment Management Plans (CAMPs)
Are surveys and reviews which assess the degree of threat to a group of taxa, or the taxa of a particular region. The need for research or PHVAs is established during this process, as are priorities for resource allocation

Global Captive Action Recommendations (GCARs)
Summarize the captive status and captive breeding priorities for recommended animals based on data collected during the CAMP process

Global Animal Survival Plans (GASPs)
Emphasize the linkage between captive programmes and programmes for conservation in the wild

Genome Resource Banking (GRB)
The approach to systematic storage and cataloguing of frozen genetic resources. An action plan for a Tiger Genome Resource Bank has already been drafted

Source: CBSG (1995)

13.3   ZOOS AND THEIR ROLE

Because of the diversity of size and type of animal collections it is difficult to say how many zoos there are, but over 1000 establishments participate in national, regional or international zoo federations. There are perhaps ten times that number of non-federated institutions which display wild animals to the public. These include such places as aviaries, bird parks, waterfowl parks, parrot gardens, reptile parks, vivaria, aquaria, butterfly houses and dolphinaria.

The numbers of people visiting all these places is unknown but the 1000 or so federated zoos receive at least 600 million visitors a year. Most of them are in the developed world (Table 31.1).

The Convention on Biological Diversity (1992) recognizes the value of *ex situ* conservation and calls for a strengthening of the conservation role of zoos. Accordingly, the World Zoo Conservation Strategy (referred to as 'The Strategy' for the rest of this chapter) has three main thrusts (IUDZG/CBSG 1993):

- direct support to programmes for the *in situ* and *ex situ* conservation of endangered species

Table 13.1 World distribution of zoos and zoo associations (source: IUDZG/CBSG 1993)

Region and estimated annual no. of zoo visitors	Regional association	National associations
North America 106 million	American Association of Zoological Parks and Aquariums (AAZPA)	Canada
Latin America 61 million	Association of Meso American Zoos (AMAZOO)	Brazil, Columbia Venezuela, Mexico, Guatemala
Europe 125 million	European Association of Zoos and Aquaria (EASA)	British Isles, Sweden, Denmark, Netherlands, Germany (plus Austria and Switzerland), Hungary, Czech/Slovak Republics, Poland, France, Spain, Italy
Africa 15 million	Pan African Association of Zoological Gardens, Aquariums and Botanical Gardens (PAAZAB)	
Asia 308 million	South East Asian Zoo Association (SEAZA)	Pakistan, India, China, Japan, Indonesia, Thailand
Australasia 6 million	Australian Regional Association of Zoological Parks and Aquaria (ARAZPA)	

- helping to increase scientific knowledge that will benefit conservation
- promoting public and political awareness of the need for conservation

13.3.1 The global zoo network

The most common network for zoo collaboration is organized at the national level; nearly 30 nations now having national zoo federations – such as the Federation of Zoological Gardens of Great Britain and Ireland. In turn, the national networks are joined through supranational or regional organizations; that for Europe being the European Association of Zoos and Aquaria (EAZA). All three levels (zoo, national, regional) are represented by a single global body – the World Zoo Organization (IUDZG, the International Union of Directors of Zoological Gardens).

The networking and organizing of zoos in this way (Table 13.1) facilitates the exchange of information, technology and live animals. It also, incidentally, adds to the plethora of acronyms and initials used in the zoo world.

13.3.2 Animal collections

A number of endangered species, including such high-profile species as the Siberian tiger, are now more numerous in captivity than they are in the wild. Many of the animals in networked zoos are recorded in a database known as ISIS (International Species Information Service), which is based in Minneapolis, USA. In 1994, its twentieth year, ISIS contained details of 213 000 living animals of 8000 taxa, along with 600 000 of their ancestors. All the live animals were housed in fewer than 500 zoos. The estimated numbers of vertebrates in all the networked zoos are shown in Table 13.2.

Table 13.2 Estimated numbers and distribution of vertebrates in zoos (source: IUDZG/CBSG 1993)

	Mammals	Birds	Reptiles	Amphibians	Fish
North America	60 000	70 000	25 000	5 000	100 000
Latin America	10 000	25 000	5 000	1 000	25 000
Europe	90 000	130 000	20 000	8 000	180 000
Asia	75 000	100 000	20 000	10 000	50 000
Africa	7 500	15 000	2 500	500	5 000
Australia	7 500	10 000	2 500	500	20 000
Totals	250 000	350 000	75 000	25 000	300 000

Estimated world total of zoo vertebrates: approximately 1 000 000.

The Strategy requires that zoos devote more space to endangered species in coordinated, managed programmes. This is beginning to happen, but not as fast as numbers would suggest because more species are also being reclassified as endangered.

Scientific and veterinary advances have increased both the welfare and longevity of zoo animals so that most animals in zoos now live longer than their wild counterparts. In addition, more species are being bred in captivity than ever before, though mammals generally do better in this regard than the other vertebrate groups. Because of these advances, and the exchange of animals within zoo networks, zoos are becoming more independent of imports of animals from the wild. But it would be contrary to the aims of conservation to break the link entirely and create separate wild and captive gene pools.

13.3.3 Programmes for species

Within the zoo network there are collaborative programmes for managing the different groups of animals. The zoo network of the British Isles, for example, has a Joint Management of Species Committee (JMSC) which has been overseeing Joint Management of Species Programmes (JMSPs) since 1992 (Lees 1994).

At the level of individual species, the management programmes are controlled by Taxon Advisory Groups (TAGs). In the British Isles there are about 40 TAGs. Each TAG has a chairperson and a working group which will include, for each target species, a species coordinator and studbook keeper (possibly the same person). The target or priority species are selected by the working groups in accordance with the CBSG Global Action Recommendations (GCARs). Most of the species programmes in the British Isles are also part of the European Endangered Species Programme (EEP) which comes under the European Association (EAZA).

13.3.4 Regional breeding programmes

At present there are regional breeding programmes for more than 300 endangered species. Each programme policy may require such measures as the exchange of animals, modifications to enclosures or husbandry practices, special attention to certain genetic lines or control of the total captive population size. The policy is revised each year and participating zoos must agree to follow the programme as closely as possible in a commitment that may involve spending money to meet the requirements.

13.3.5 Zoo capacity and the selection of species

If the 1000 or so networked zoos each have an average of 500 spaces for larger vertebrates in breeding programmes, and this is a reasonable assumption, then about half a million animals could be housed; and with 500 specimens in each breeding programme then 1000 breeding programmes could be catered for. This number might be doubled if all the small zoos, not yet in networks, were to cooperate. However, The Strategy does not aim to accommodate fully viable captive populations of any but the most critically endangered species. So-called nucleus populations of about 100 individuals might suffice as an *ex situ* backup for species whose future in the wild looks rather more secure.

The Strategy calls for selection of species to be based on conservation needs as revealed by assessments and analyses (CAMPs and PHVAs) from the field, and with due regard to the strengths and weaknesses of particular zoos. The ultimate aim is to produce regional collection plans (RCPs) that will contribute to a Global Captive Action Plan (GCAP) for each taxon in a breeding programme.

13.3.6   Biotechnology in zoos

A variety of artificial reproduction techniques have been developed during the past decade or so, largely from research on human reproduction and contraception (Holt 1994). It is easier to transport sperm and embryos than to move animals. Reproductive technologies can also be used to manipulate sex ratios, litter size or the incidence of twinning. This can help to speed up population growth in critically endangered species.

There are further advantages in being able to store sperm, eggs and embryos by freezing. The idea of a 'frozen zoo' would have been science fiction not very long ago but the European Community is now funding a research programme to determine the real-life possibilities (Seymour 1994). Already, sperm which has been stored by freezing has been used in the successful reproduction of antelopes, deer, apes and wolves. In addition to humans and some domestic animals, baboons, marmosets and eland have been produced from embryos stored by freezing (IUDZG/CBSG 1993).

13.3.7   Conservation of behaviour

Genetically determined behaviour can be transmitted through a great many generations of animals which have no opportunity to express that behaviour, but learned behaviour cannot be acquired through the genes. Learned knowledge and skills can only come from experience, although there may be an inherited propensity to learn. The greater the capacity for learning by an animal, the more important learned behaviour is likely to be for its survival and social functioning, but even captive-reared molluscs can show behavioural 'naïvety' compared with their wild counterparts (Lucas, Chapter 5, this volume). An orang-utan spends six years learning from its parents and other group members. Some animals, such as birds of prey and other predators, may not learn much directly from their parents but parental support can be essential for the young until they learn survival skills by trial and error.

Captive animals deprived of the opportunity to express normal behaviour may show abnormal and often stereotyped behaviour. Devising ways of enriching the zoo environment so as to minimize this problem is now standard practice in good zoos. Making animals search for food and deal with it in ways that stimulate natural foraging, for example, is likely to reduce the time they spend performing useless, stereotyped movements (Shepherdson 1994). This is not only a welfare issue; it has clear implications for conservation biology when the aim is to produce normal offspring through captive breeding and possibly to release them into the wild.

There is evidence that reproduction itself can be adversely affected not only by too much disturbance but also by the stress of under-stimulation in captivity. An environment enriched to the point of including short periods of stress from threatening events may be beneficial for reproduction in some animals (Moodie and Chamove 1990, cited in Shepherdson 1994). It seems that the occasional need for an adrenalin rush as a 'turn on' is not unique to young humans.

The Strategy urges zoo workers to ensure that social animals are kept in groups which permit natural behavioural needs and responses, and to construct and furnish enclosures so that a wide range of natural behaviours can be expressed.

13.4 WILD AND CAPTIVE

In Chapter 2 reference was made to the conservation problems presented by small, fragmented populations in the wild. Zoo populations can be regarded in the same way and The Strategy calls for zoo populations to be considered alongside small relict populations in the wild, as components of the same metapopulation of the species. The intention is to try to manage *in situ* and *ex situ* groupings of endangered species interactively. They do present common management problems and under some circumstances the exchange of genetic material between zoos and relict wild populations could increase the survival chances of both groups. Well managed zoo populations are probably less susceptible to genetic degeneration than are relict populations of comparable size left under natural conditions; zoo populations can also benefit from genetic inputs from the wild.

In zoos it is thought to be possible to maintain over 90% of the original genetic variability for the long term (200 years or more) by following the management guidelines put forward by The Strategy. They include:

- trying to begin with a founder group of at least a few dozen animals
- increasing the population size from the founders as quickly as possible to at least 250–500 animals (as a general rule)

- for effective genetic mixing, keeping sex ratios as close as possible to 1 : 1, or trying to effect this by such practices as exchanging males in polygamous groups
- avoiding inbreeding as far as possible
- as soon as target population size is reached, extending the generation time so that animals reproduce when they are older and the potential rate of genetic change is slowed
- trying to evade unnatural selection pressures in the zoo environment
- trying to add a small number of unrelated animals to the population in each generation

13.4.1   Zoo populations and genetics

The question as to how many specimens constitute a viable population for the long term has no simple answer because the minimum viable population (MVP) for one species may have little relevance for another. A population and habitat viability assessment (PHVA) requires inputs under four headings:

- biological characteristics of the species (the phenotype)
- environmental assessments (quality, area, security)
- demography and population fitness (age distribution, sex ratios, growth rates)
- genetic make-up (the genotype)

The relationship of a population to the metapopulation (if any) is also important.

The zoo environment is more predictable than the wild so the element of chance is reduced to some extent. Demographic manipulations might also be possible so as to maximize the breeding potential. This increases the relative importance of genetic considerations in the management of zoo populations. In the wild, as we have seen, habitat security is usually the major concern.

Genes occur in paired form at each locus on a chromosome. The two elements of a pair are called **alleles**; one is contributed by the mother and the other by the father. Vertebrate chromosomes contain around 100 000 loci and for a particular locus there may be only one kind of allele or there may be different ones which can occur in combination. For example: if the gene pool of a population contains two alleles (A_1 and A_2) for locus A, then any individual in the population can have one of three possible combinations: A_1A_1 or A_1A_2 or A_2A_2. An individual with the first and third combination is **homozygous** at that locus and an individual with the second combination is **heterozygous**. The conservation aim is to maintain **genetic variance** and this correlates very well with heterozygosity – the proportion of loci that are heterozygous in an average individual (Caughley and Sinclair 1994).

13.4.2   Inbreeding

Continued breeding from small populations (which remain small) will lead to a decline in heterozygosity. Alleles may not be passed on in the unequal reproduction of individuals and so are lost by chance. Also, inbreeding will tend to bring similar alleles together because genetically close parents are sharing more genes than non-related parents. Inbred individuals therefore have a higher probability of being homozygous for a parent's genes and may show signs of inbreeding depression as a result.

The usual explanation for inbreeding depression is that inbreeding brings together deleterious recessive alleles. The gene pools of most populations contain many of these sublethal recessives but their effects are not expressed in the phenotype when they are combined at their locus with a dominant allele. Thus, if the A_1 were recessive and harmful, but its effects masked by the dominant A_2, then only the homozygous individuals, A_1A_1, would be disadvantaged. Different taxa show different sensitivities to inbreeding depression. In a range of mammals there was a 33% higher mortality, on average, among the young of parent–offspring and full sibling matings than among those of unrelated parents (Ralls *et al.* 1988). Captive populations of the Bali starling (*Leucopsar rothschildi*) have shown lower fertility and higher mortality attributable to inbreeding (van Balen and Gepak 1994).

It would be expected that natural selection would have removed deleterious genes from populations which have a long history of inbreeding in the wild. Such populations ought to be comparatively resistant to further inbreeding depression but this was not confirmed by studies on mice (Lacy 1992). Alternatively, irrespective of the masking of deleterious recessives, it is possible for a heterozygote to be fitter in some way than either homozygote and perhaps there may be advantages in general heterosis. In view of all this uncertainty, the wisest course for zoos is to try to avoid inbreeding.

Since it is impossible to say exactly how many individuals are needed to make up a viable populations, conservation biologists have had to make do with very rough rules of thumb. On the basis of laboratory work with fruit flies and some wider observations, it has been suggested that at least 500 individuals are needed to avoid long-term loss of genetic variability and that at least 50 individuals are needed to limit the effects of inbreeding in the short term (Franklin 1980; Soulé 1980). However, these numbers are only to be regarded as an order of magnitude (Craig 1994) and it would certainly be wrong to abandon hope for a species because its numbers have fallen below these arbitrary levels. There have been apparent recoveries from smaller numbers (e.g. Mauritius kestrel, see section 9.7.5). While only marginally relevant, it is worth a mention in passing that all the captive golden hamsters (*Mesocricetus auratus*) in the world are believed to have originated from a single pregnant female discovered in the 1930s.

13.4.3   Reintroductions

Within The Strategy the ultimate goal of *ex situ* breeding programmes is to support the survival of endangered species in their natural environments. Zoos are doing this through education, research and project promotion. Biotechnology, in the future, may also prove to be very helpful for endangered populations *in situ* as well as *ex situ*. Restocking and reintroducing animals to the wild, however, is much more problematical.

The problems that caused the species to be endangered in the first place must obviously be addressed and this is already the main thrust of conservation efforts that do not involve zoos. Zoos can provide suitable animals for reintroduction as and when safe natural habitat is restored, but it is not clear how zoos can be especially helpful in securing natural habitat. Indeed, historically, the great majority of reintroductions have been undertaken not by zoos but by the state and federal wildlife authorities of industrialized countries. An analysis of the reintroductions record led Beck *et al.* (1994) to conclude that some zoos and zoo critics exaggerate the importance of reintroduction as a zoo conservation function. Once again, the question of time-scale arises; while there may be no chance of reintroducing an animal to the wild in the short term, there will never be a chance if it is allowed to become extinct.

The Beck analysis was based on the records of 145 reintroduction attempts this century. The projects involved captive-born animals of 126 vertebrate species. The authors defined successful projects as those in which the wild population subsequently reached at least 500 individuals, were not being provisioned or otherwise supported by human intervention, and were expected to be self-sustaining on the basis of a formal analysis. On these criteria there was evidence that only 11% were successful, although that is not to say that 89% of the projects have definitely failed or will fail. In another study, which allowed managers of reintroduction projects to judge their own successes, it was estimated that 38% of projects involving captive-bred animals were successful (Griffith *et al.* 1989).

Beck *et al.* (1994) found roughly equal proportions of successes among mammals, birds and reptiles/amphibians. There were no successful reintroductions of fish or invertebrates. In looking for correlates of success, the authors found that successful projects had more frequently provided local employment and community education programmes, and the projects had continued over a number of years. These preliminary findings highlight the importance of socio-economic criteria and long-term funding; it is not enough to have biologically sound release techniques. This point is clearly reflected in the Guidelines for Reintroductions produced by the IUCN Reintroduction Specialist Group (IUCN/SSC/RSG 1995).

13.5 PRIVATE COLLECTIONS

As a top priority zoos must avoid losing public support and going broke. There is clearly a limit to the proportion of a zoo's resources that can be devoted to species with low exhibition value. This includes a great many small and superficially similar vertebrates among fish, frogs, lizards, snakes and small birds. These are the sorts of animals which are kept in very large numbers by specialist collectors, hobbyists and casual pet-keepers.

13.5.1 The live animal trade

Most international trade in live animals is in ornamental fish (500–600 million freshwater and marine); reptiles (5–6 million); birds (2–5 million) and primates (25 000–30 000). However, more than 95% of the freshwater fish are captive bred and possibly more than half the reptiles are farm-bred freshwater turtles (terrapins) (Hemley 1994). Of all birds passerines of the finch type (families: Estrildidae, Fringillidae, Ploceidae) are traded in the largest numbers, with most originating from Africa. Birds of the parrot group (order Psittaciformes) are the next largest group in trade. Most birds of all kinds are destined for the EU, the USA and Singapore (Mulliken *et al.* 1992).

From a conservation perspective the live animal trade has been, and still is, a serious drain on some wild populations and has contributed to the decline of certain species, notably in the parrot group (Iñigo-Elias and Ramos 1991; Thomsen and Brautigam 1991). It is also the cause of serious welfare concerns, with heavy mortalities during capture, storage and transportation (Swanson 1992).

People seem to be less concerned for the welfare of reptiles and amphibians though there may be no less cause. A significant proportion of the trade in live animals of these groups is not for the pet trade but for food and traditional medicine. For these purposes up to 300 tonnes of tortoises and freshwater turtles are traded annually between East Asian countries, raising both conservation and welfare concerns (Oryx 1996). The international trade in frogs' legs for cooking also involves significant trade in live frogs; France imported about 500 tonnes of live frogs in 1986 (Le Serrec 1988).

In the USA there are signs that the live reptile trade is increasing as trade in wild birds declines. The US Wild Bird Conservation Act of 1992 sharply reduced the volume of imported wild birds – especially parrots (Hemley 1994). Welfare groups have also pressured some airlines into refusing to carry live bird consignments (Edwards and Thomsen 1992). Increasing production from captive breeding in consumer countries seems set to make further inroads into the wild bird trade (Mulliken *et al.* 1992).

These recent moves against the wild bird trade bring the concept of conservation through utilization very sharply into focus. The trade has been justifiably opposed on genuine conservation grounds as well as by welfare groups. It is easy to oppose but it can be far more difficult to introduce constructive alternatives. Edwards and Thomsen (1992) have considered the possibilities and concluded that eliminating the trade will do little for conservation because most species in trade are most seriously threatened by habitat loss. Nor will captive breeding in the consumer countries encourage conservation in the producer countries. Yet the major producer countries not only lack the technical capacity for captive breeding, but commercial attempts would again raise welfare concerns. In most of the exporting countries the concept of animal welfare with respect to wild species is not well established, and animal welfare laws covering wild species are non-existent or not enforced.

Government officials of the exporting countries cited the habitat protection value of the wild bird trade as their main reason for supporting it, but no government was actually using that incentive to forestall land conversion.

Edwards and Thomsen (1992) emphasized, among other things, the need for technical assistance to exporting countries in order to help develop proper welfare standards and captive breeding programmes, and to explore the potential for ranching wild birds in the natural environment.

13.5.2   Pet shops

Although some pet shops do carry significant stocks of exotics, the majority of animals which one sees in such shops are domesticated and so have little relevance for wildlife conservation. Domestic cats should be mentioned here as a special case because they kill such great numbers of wild creatures. Every year tens of millions of birds are killed by cats in Britain alone.

The main concern for most animals sold from pet shops is for their welfare. But whereas pets of all kinds can end up in unsuitable homes, or with owners who soon lose interest, the problem of owner ignorance is likely to be greatest where exotic, non-domestic animals are involved. Welfare and conservation concerns therefore come together over this segment of the pet-shop trade.

13.5.3   Amateur specialists

It is notoriously difficult to acquire accurate figures for international trade in animals but it is even more difficult to find out what happens to them after they have passed through the importation procedures of the destination country. It seems reasonable to suppose that only the most popular species will be destined mainly for pet shops while others will be held by dealers and sold from catalogues to collectors and other specialized buyers.

These customers include individuals who have built up a great deal of specialist knowledge and skill and can be regarded as expert in keeping and breeding particular taxa. They are the sorts of enthusiasts who belong to societies and subscribe to journals which cater for their special interests – birds, reptiles and amphibia, or fish.

In aviculture, for instance, 'much husbandry skill and expertise as well as the great majority of parrots, are found within the private sector' (Wilkinson 1992). Similar observations could be made about other groups. Details of nest-building in the Australian firetail finch (*Eblema bella*) were first recorded by a private enthusiast who sat observing his aviary for 6–7 hours a day for a total of 295 hours (Shephard 1989).

Many hobbyists are involved with taxa or hybrids that are irrelevant for conservation and it would be easy to overstate the conservation case. But when private individuals are able to make conservation inputs, their involvement is likely to be through low profile projects that would not have been funded from public money or allowed to overload the resources of zoos. They are all the more useful for that.

13.5.4 Interaction and coordination

The amateur and professional sectors need not remain entirely separate but whereas collaboration is common in environmental projects, the potential of the private sector appears not to have been so widely recognized or accepted where captive management is involved. And yet several reintroduction programmes for birds of prey were collaborative efforts between falconers, NGOs and national or state wildlife authorities (section 9.7.5). There are very few examples from Australia but the Commonwealth Scientific and Industrial Research Organization (CSIRO) and state wildlife authorities have worked in collaboration with private aviculturalists on some captive breeding projects; those for the plains wanderer (*Pedionomus torquatus*) and the Lord Howe woodhen (*Tricholimnas sylvestris*), for example.

Where individuals become involved in networked breeding programmes, the necessary coordination is achieved through the Taxon Advisory Groups. Membership of the parrot TAG for the British Isles, for instance, includes private breeders and representatives of specialist societies (lories, Amazons, lovebirds) and bird parks as well as mainstream zoos (NEZS 1995). Similarly, the Fish and Aquatic Invertebrate TAG (FAITAG) for the British Isles includes members of the British Cichlid Association who have begun to specialize in threatened cichlid fish from Lake Victoria. The Association provides an interface between hobbyists and professional institutions. The British Killifish Association performs a similar role for cyprinodont fish (Reid and Whitear 1994).

The Australian Reptile and Amphibian TAGs include a representative from an affiliation of the region's herpetological societies in recognition of the 'potential for involvement from reputable private herpetologists' (Banks 1993). However, Australian law is not helpful in realizing this potential for it does not encourage liaison between amateur enthusiasts and professionals (section 13.5.7).

Of course not all private enthusiasts want to be involved in coordinated breeding programmes. The Strategy requires that private breeders who participate must do so on very strict terms. All pairings of breeding programme animals must be in accordance with the master plan for the species; there must be no involvement in commercial trade; the animals must be registered in studbooks and ISIS and not disposed of without approval from the species coordinator.

13.5.5   The problem of trade

Keeping animals, even small ones, costs money and many private breeders help to pay for their hobby by selling surplus stock. Since this cannot be done with animals in breeding programmes, other, saleable, species may have to be used to subsidize a private breeder's participation in a programme. The Strategy endorses a policy of 'nil commercial value' for zoo animals and calls for zoo associations to eliminate price-tags and dissociate themselves from commercial trade.

It is right and proper that zoos should not be supporting unscrupulous animal dealers and exploitative trade, but a policy of nil commercial value for wildlife is clearly at odds with the principle of conservation through utilization. Demands for wildlife to pay its way in the wild cannot be met with commercially worthless animals. Nor is there any conservation advantage in preventing an honest private breeder from selling surplus stock to another honest keeper. Indeed, the export of one or two pairs of captive-bred platypus (*Ornithorhynchus anatinus*) to Japan could provide significant funds for platypus habitat conservation within Australia. But the sale, from a private breeder, would be illegal under Australian law (*Sydney Morning Herald*, 13 April 1996). Conflicts of this sort cannot be resolved by policy-makers who are bound by philosophical dogma as to what is 'right' or 'wrong' with regard to wildlife and monetary value. A pragmatic approach is needed to tackle the increasingly complex problems of trade and conservation.

It is probably rare for private collectors to entirely cover the costs of their hobby by selling surplus stock, but some do become fully commercial by opening to the public and charging an admission fee.

13.5.6   Commercial specialists

Like all private collections, small specialized exhibitions may or may not have any relevance for conservation. Much depends on the interests and abilities of the owners as well as on their financial and material resources. In their survey of captive raptors in the UK, Cromie and Nicholls (1995) found that most aviaries in federated zoos provided better visitor education than did aviaries in non-federated centres. Nevertheless, some federated zoo aviaries were rated 'very poor' both in terms of educational design and the quality of interpretation provided, whereas a few aviaries in non-federated centres were rated 'good' and 'moderately good'. Many were rated 'satisfactory'. Evidently, small commercial specialists sometimes make a better job of conservation education than do some larger zoos.

Education has been combined with more direct conservation measures in one Australian enterprise. The impact of introduced predators on Australia's native fauna has long been recognized but it was a private individual who took the initiative to create a native fauna sanctuary by fencing out all cats and foxes. In 1969, John Wamsley, who used to be head of a university maths department, bought 14 hectares of degraded farmland in the Adelaide Hills of South Australia and called it Warrawong Sanctuary. He fenced out all feral predators and replanted it with native trees and shrubs. Then he reintroduced some of the native mammals that formerly lived in the Adelaide Hills. The small mammals did remarkably well in terms of population increase. For example, within about 11 years a rare and endangered member of kangaroo superfamily, the brush-tailed bettong (*Bettongia penicillata*), increased from six individuals to over 200.

In 1985 Warrawong Sanctuary was opened to the public as an ecotourism venture offering guided walks at dawn and dusk. By 1992 it had won a dozen Australian awards including the South Australian Tourism Award in the Environmental Tourism Category. Wamsley is now managing director of Earth Sanctuaries Ltd and Warrawong is one of several sanctuaries, totalling 80 000 ha, which the company is developing on a similar basis (*Sydney Morning Herald*, 13 April 1996).

The Warrawong story is unique in Australia and the combination of circumstances which led to its success may be hard to find in other countries. Enterprise and dedication, a variety of wild animals badly affected by a problem that was technically easy (if expensive) to solve, suitable and available land, an interested and commercially supportive public and relative freedom from official obstruction were all essential ingredients.

With regard to the last component, while Wamsley's activities were not prevented or officially discouraged, he has declared that government regulations so discriminate against the private sector in matters of wildlife management that they constitute the biggest problem his company has to face. Yet he has pointed out, in a statement to the Australian Federal Government, that whereas his sanctuaries have increased the numbers of at least eight rare and endangered mammals during the past 20 years, all rare and endangered mammals under official protection in eastern Australia are fewer in number than they were 20 years ago.

It will be appropriate to end this chapter by considering official policies in a wider context.

13.5.7   Conservation, captivity and the law

It was noted in Chapter 2 that the most common legislative approach to wildlife protection at national level has been to prohibit 'taking', possession and/or commercial trade in wild species; the species usually being grouped in appendices under different categories of protection. While this is intended to discourage deliberate persecution of wild animals, it is possible to exaggerate the importance of species-based legislation to the extent of diverting public attention from the more serious problem of habitat loss. Moreover, in many countries there is little or no enforcement of wildlife protection laws, while at the other extreme there are countries which administer such strict legislation as to prevent genuine enthusiasts from acquiring valuable experience and expertise. In the latter context, Australia compares unfavourably with the USA and the UK.

(a) Australia

Australia has been criticized for having become overenthusiastic about species legislation since the 1970s – during which period land clearance has continued on a vast scale. Understandably, much of the criticism has come from frustrated hobbyists who feel that bulldozers have been given legal priority of access to native fauna, but both professional biologists (e.g. Ehmann and Cogger 1985) and legal experts (e.g. Chapman 1993) have also expressed the view that legislation which is over-restrictive to genuine interests can work against conservation and be counter-productive.

John Weigel (1992) records that amateur herpetologists can, with difficulty, obtain permits to acquire certain common reptiles from other captive sources but in some states they are not allowed to breed them! Australian aviculturalists are forbidden to keep birds of prey, whether they intend to breed them or not.

Most wildlife professionals were probably keen amateurs to begin with. One Australian herpetologist, writing with some prescience more than 30 years ago, cautioned against discouraging young amateurs before their interest could mature. It was important, he wrote, 'that the growing interest of Australians in their own fauna, reptiles especially, should not be soured by bureaucracy' (Worrel 1966). A young person who was prosecuted for keeping a common and harmless snake as a pet, when his neighbours could have killed it with virtually no risk of prosecution, might indeed feel sour. Some of the more unpleasant relationships between hobbyists and wildlife authorities in Australia have been documented by Hoser (1993).

(b) United Kingdom

Under the Wildlife and Countryside Act (1981) all birds are protected but there are extensive provisions for permitting specified actions under licence – including capture for approved breeding programmes – subject to strict criteria. Most common mammals may be kept and it is not illegal for amateur herpetologists to collect specified common reptiles (e.g. grass snake, adder, smooth newt) for captive breeding, but a licence would be needed to sell them.

(c) USA

Each American state has its own wildlife protection agency and state regulations differ. At the federal level the Endangered Species Act is the primary legal mechanism for wildlife protection. However, only species classified as 'endangered' or 'threatened' fall under its provisions. Most of the animals so classified are mammals and birds; only very small percentages of reptiles and amphibians are covered by the Act. Endangered and threatened species may be taken, kept or traded only under a special permit which is unlikely to be granted to a private enthusiast, but there is provision for captive-bred raptors to be kept and traded (section 9.6.1).

Species not covered by the Act and not otherwise protected may be taken and freely traded. The importation of exotic species is subject to state and federal laws (in addition to CITES) but transactions in unprotected domestic reptiles are mostly uncontrolled. In general, it appears that while government does not actually encourage hobbyists to keep wild animals, it nevertheless leaves them ample opportunity to pursue their interests and develop a high level of expertise.

ACKNOWLEDGEMENTS

I owe special thanks to Tomme Young for her help from the USA, and am particularly grateful to Gordon McGregor Reid and Roger Wilkinson of Chester Zoo, who revealed for me the inner workings of TAGs. Others who provided information include Steve Gibson, John Holmes, John Weigel, Raymond Hoser, John Wamsley, Caroline Lees, Nigel Steele-Boyce and Chris Banks. I thank them all.

REFERENCES

Banks, C.B. (1993) A regional approach to managing reptiles and amphibians in Australasian zoos, in *Herpetology in Australia: a Diverse Discipline* (eds D. Lunney and D. Ayers). Transactions of the Royal Zoological Society, Mosman, New South Wales, pp. 59–65.

Beck, B.B., Rapaport, L.G., Stanley Price, M.R. and Wilson, A.C. (1994) Reintroduction of captive-born animals, in *Creative Conservation: Interactive*

Management of Wild and Captive Animals (eds P.J.S. Olney, G.M. Mace and A.T.C. Feistner). Chapman & Hall, London, pp. 265–86.

Caughley, G. and Sinclair, A.R.E. (1994) *Wildlife Ecology and Management*. Blackwell, Cambridge, Mass.

CBSG (1995) *Introducing the Captive Breeding Specialist Group*. CBSG, Minnesota.

Chapman, C. (1993) Preface to *Smuggled: The Underground Trade in Australia's Wildlife* (R. Hoser). Apollo Books, Sydney.

Craig, J.L. (1994) Metapopulations: is management as flexible as nature?, in *Creative Conservation: Interactive Management of Wild and Captive Animals* (eds P.J.S. Olney, G.M. Mace and A.T.C. Feistner). Chapman & Hall, London, pp. 50–66.

Cromie, R. and Nicholls, M. (1995) The Welfare and Conservation Aspects of Keeping Birds of Prey in Captivity. Report to the RSPCA, London.

Edwards, S.R. and Thomsen, J.B. (1992) A management framework for the wild bird trade, in *Perceptions, Conservation and Management of Wild Birds in Trade*. Traffic International, Cambridge, pp. 151–65.

Ehmann, H. and Cogger, H.G. (1985) Australia's endangered herpetofauna: a review of criteria and policies, in *Biology of Australasian Frogs and Reptiles* (eds G. Grigg, R. Shine and H. Ehmann). Surrey Beatty, Chipping Norton, NSW, pp. 435–47.

Franklin, I.R. (1980) Evolutionary change in small populations, in *Conservation Biology: An Evolutionary–Ecological Perspective* (eds M.E. Soulé and B.A. Wilcox). Sinauer Associates, Sunderland, Mass, pp. 135–49.

Griffith, B., Scott, J.M., Carpenter, J.W. and Reed, C. (1989) Translocation as a species conservation tool: status and strategy. *Science*, **245**, 477–80.

Hemley, G. (ed.) (1994) *International Wildlife Trade: A CITES Sourcebook*. Island Press, Washington, DC.

Holt, W.V. (1994) Reproductive Technologies, in *Creative Conservation: Interactive Management of Wild and Captive Animals* (eds P.J.S. Olney, G.M. Mace and A.T.C. Feistner). Chapman & Hall, London, pp. 144–66.

Hoser, R. (1993) *Smuggled: The Underground Trade in Australia's Wildlife*. Apollo Books, Sydney.

Iñigo-Elias, E.N. and Ramos, M.A. (1991) The psittacine trade in Mexico, in *Neotropical Wildlife Use and Conservation* (eds J.G. Robinson and K.H. Redford). University of Chicago Press, Chicago, pp. 359–79.

IUCN/SSC/RSG (1995) *Guidelines for Reintroductions*. Reintroduction Specialist Group, Nairobi.

IUDZG/CBSG (1993) *The World Zoo Conservation Strategy*. Chicago Zoological Society, Brookfield, Illinois.

Lacy, R.C. (1992) The effects of inbreeding on isolated populations: are minimum viable population sizes predictable?, in *Conservation Biology: The Theory and Practice of Nature Conservation Preservation and Management* (eds P.L. Fiedler and S.K. Jain). Chapman & Hall, London, pp. 277–96.

Lees, C. (ed.) (1994) *Annual Report of the Joint Management of Species Programmes*. Federation of the Zoological Gardens of Great Britain and Ireland, London.

Le Serrec, G. (1988) France's frog consumption. *TRAFFIC Bulletin*, **10**, (1/2), 17.

Moodie, E.M. and Chamove, A.S. (1990) Brief threatening events are beneficial for captive tamarins. *Zoo Biology*, **9**, 275–86.

Mulliken, T.A., Broad, S.R. and Thomsen, J.B. (1992) The wild bird trade – an overview, in *Perceptions, Conservation and Management of Wild Birds in Trade.* Traffic International, Cambridge, pp. 1–41.

NEZS (1995) *JMSP Parrot Taxon Advisory Group Annual Report, 1994.* North of England Zoological Society, Chester.

Oryx (1996) Threat to tortoises and freshwater turtles, *Oryx*, **30** (2), 97. (Editor's report from *TRAFFIC Despatches*, Sept. 1995.)

Ralls, K., Ballou, J.D. and Templeton, A. (1988) Estimates of lethal equivalents and the cost of inbreeding in mammals. *Conservation Biology*, **2**, 185–93.

Reid, G.M. and Whitear, J. (eds) (1994) *Report of the Fish and Aquatic Invertebrate Taxon Advisory Group, 1994, with Proceedings of the 1944 Fish and Aquatic Invertebrate Conservation Workshop*, North of England Zoological Society, Chester.

Seymour, J. (1994) Freezing time at the zoo. *New Scientist*, **141** (1910), 29 January, 21–3.

Shephard, M. (1989) *Aviculture in Australia.* Reed Books, Chatswood, New South Wales.

Shepherdson, D. (1994) The role of environmental enrichment in the captive breeding and reintroduction of endangered species, in *Creative Conservation: Interactive Management of Wild and Captive Animals* (eds P.J.S. Olney, G.M. Mace and A.T.C. Feistner). Chapman & Hall, London, pp. 167–77.

Soulé, M.E. (1980) Thresholds for survival: maintaining fitness and evolutionary potential, in *Conservation Biology: An Evolutionary–Ecological Perspective* (eds M.E. Soulé and B.A. Wilcox). Sinauer Associates, Sunderland, Mass, pp. 151–69.

Swanson, T.M. (1992) Economics and animal welfare, in *Perceptions, Conservation and Management of Wild Birds in Trade.* Traffic International, Cambridge, pp. 43–57.

Thomsen, J.B. and Brautigam, A. (1991) Sustainable use of neotropical parrots, in *Neotropical Wildlife Use and Conservation* (eds J.G. Robinson and K.H. Redford). University of Chicago Press, Chicago, pp. 359–79.

van Balen, B. and Gepak, V.H. (1994) The captive breeding and conservation programme of the Bali starling (*Leucopsar rothschildi*), in *Creative Conservation: Interactive Management of Wild and Captive Animals* (eds P.J.S. Olney, G.M. Mace and A.T.C. Feistner). Chapman & Hall, London, pp. 420–30.

Weigel, J. (1992) How Australian wildlife protection policies impede herpetology. *Herpetofauna*, **22** (2), 7–14.

Wilkinson, R. (1992) Parrot breeding programmes in public zoos and private aviculture: some problems and perspectives. *Proceedings of the Avicultural and Veterinary Conference*, London Zoological Society, London, pp. 97–102.

Worrel, E. (1966) The unpopular ones, in *The Great Extermination* (ed. A.J. Marshall). Heinemann, Melbourne, pp. 75–94.

14

Loving them and leaving them: wildlife and ecotourism

14.1 INTRODUCTION

Tourism is an enormous industry and still one of the fastest growing industrial sectors. In 1950 about 25 million tourists generated US$8 billion. In 1991 about 450 million tourists generated about US$260 billion. International tourism arrivals are expected to double again between 1991 and the year 2010 (WTO 1992). Countryside pursuits are also booming in popularity. More than 70% of Americans now participate in 'rural recreation' and the participation rate in other OECD countries is only slightly lower (OECD 1994). This combination of interest in travel and the outdoors is causing the ecotourism subsector to grow at twice the average tourism rate (Cater 1994).

So what is ecotourism? The many definitions in print appear to have four elements in common:

- the natural environment
- ecological and cultural sustainability
- education and interpretation
- economic benefits at the local level

No doubt everyone would agree that ecotourism is nature-based tourism but different writers have placed different emphases on local culture, sustainability, education, and local economic benefits. Karen Ziffer (1989), cited in Giannecchini (1993), restricts the title of 'ecotourist' to one who

> visits relatively undeveloped areas in the spirit of appreciation, participation and sensitivity. The ecotourist practices a non-consumptive use of wildlife and natural resources and contributes to the visited area through labour and financial means aimed at directly benefitting the conservation of the site and the economic well-being of the local residents.

Conservation and the Use of Wildlife Resources. Edited by M. Bolton.
Published in 1997 by Chapman & Hall. ISBN 0 412 71350 0.

In contrast, Donald Hawkins (1991), also cited in Giannecchini (1993), considers that 20–25% of leisure travel could be defined as nature tourism or broadly defined as ecotourism. Many in the tourism industry are not concerned with definitions at all; for them the prefix 'eco' is nothing more than useful marketing babble (Cater 1994).

14.2   TRENDS AND PHASES

Nature-based tourism, regardless of how it is defined (eco, green, responsible, alternative) is now being strongly promoted on every continent and in every conceivable form. There is a demand for increasingly remote destinations and for more active, penetrative, personalized holidays as an alternative to package tours. New destinations may be pioneered by biologists and travel writers, followed by adventurous back-packers, but after a time it can be expected that comfortable facilities will be built and the next phase will be one in which small numbers of wealthy people arrive for exclusive holidays. This second phase had been reached in the Royal Chitwan National Park of Nepal during the 1970s. At that time back-packers could stay in cheap accommodation close to the park while inside the park the Tiger Tops Jungle Lodge alone catered for small groups who arrived by chartered aircraft. From the airstrip the lodge guests were transported to the lodge in grand style on the backs of elephants.

Phase three of development can be considered as that stage of development which fulfils the needs of middle income people. At Chitwan there are now seven lodges inside the park and more than 40 cheaper lodges and other forms of accommodation outside. Ultimately, where a destination has the potential for mass tourism, package tours will be introduced and the destination will have run its course and reached phase four. By then, those individuals seeking unique experiences in remote places will be going elsewhere. The progression can occur at the scale of single sites or whole regions. Alaska is set to move from the phase of discovery and exclusivity to one of rapid tourism growth (Prosser 1994).

14.3   THE FORCES OF ECOTOURISM

Tourism is a highly fragmented and competitive industry, not subject to overall control. Nor is the tourism market able to control its own path of development so as to best serve its own long-term interests.

14.3.1   Who is involved?

The stakeholders in ecotourism have been described by Durst (1994) under the following headings:

- tourists (with different desires and expectations)
- local landowners and users (potential winners and losers at the site)
- government agencies
  - tourism authorities (often small and underfunded)
  - natural resource management agencies (also often underfunded and understaffed)
  - agencies responsible for infrastructure (roads, electricity, water and sewerage, etc.)
  - planning and finance ministries (with interests at the national level)
- tour operators (with direct influence on conduct of ecotourism and interactions with locals)
- investors and concessionaires (private investors in lodges, transportation services, artisan shops, restaurants, etc. can be most efficient but operations need to be regulated by governments and local residents)
- non-government organizations (many international NGOs are active in ecotourism, including conservation organizations, scientific and educational organizations, and one or two ecotourism bodies)
- development assistance agencies and donors (especially in planning, training and funding, often influenced by NGOs)

14.3.2 Where does the money go?

The economics of ecotourism can be considered at the level of the private investors, the national economy, the local economy, and the budget of the protected area. The benefits are rarely well distributed through the four levels and the revenues usually decrease rapidly in the order listed (Groom *et al.* 1991). In fact, the largest single item of a tourist's expenditure is often the airline ticket so unless the destination country's airline is used, the bulk of the economic benefit may not even flow to the host nation. Use of package tours, where the package includes accommodation, increases the proportion of a tourist's expenditure which is captured at the transnational level. The economic benefits and costs of PAs could therefore be considered at five levels but very little empirical work has been done on the distribution of PA costs and benefits (Wells 1992).

Barnes *et al.* (1992) examined the value of tourism in Khao Yai National Park, which contains most of the remaining forest in north-eastern Thailand. Ninety-five per cent of visitors to the park are Thai nationals so most of the expenditure occurs within the country. The Tourist Authority of Thailand (TAT) operates lodges, restaurants, a golf course, souvenir shops and other facilities, the profits from which accrue directly to the TAT. Only a small percentage of tourist expenditure accrues to the park through entrance fees and accommodation charges. Surveys indicated that both foreigners and Thai nationals would be prepared to pay more to enter the park than is actually charged. This difference between what consumers are willing to pay and what they actually pay is known as the consumer surplus.

It is obviously impracticable to charge for entrance to PAs according to individual means so developing countries must decide what is an appropriate disparity between residents and non-residents. In Kenya it was estimated that the Lake Nakuru National Park returned only about 5–10% of the recreational value of the park to the Kenya Wildlife Service. Ninety per cent of this was from entrance fees and it was estimated that these amounted to less than 0.1% of a foreign tourist's total expenditure on the visit to Nakuru. Accordingly, on the assumption of a large consumer surplus, entrance fees for non-residents were increased in 1991 by 310% – bringing them to the equivalent of about US$13, compared with US$1 for residents (Navrud and Mungatana 1994).

While the consumer surplus is a legitimate source of revenue to be tapped at the ecotourism site, more fundamental measures are needed to direct a greater proportion of tourism benefits to the local and protected area economies. There is no simple solution, for, as Wells (1992) has observed: 'The underlying causes of cost–benefit imbalances at local, national and global scales are extremely difficult to untangle.'

14.3.3 Valuing animals for tourism

Whereas recreational hunting can earn more per animal than commercial hunting, tourism, in effect, sells the same animals repeatedly so that the most favoured species can be extraordinarily valuable. Economists can estimate direct use value of a natural attraction in two ways: the travel cost method (what people would be prepared to pay to visit the attraction) and the contingent valuation method; based on asking people what they would be prepared to pay in order to maintain, increase, prevent losses etc. to the attraction. (Contingent valuation procedures can be adapted for such purposes as assessing environmental damage from toxic spills.) By both these methods the viewing value of Kenya's elephants has been estimated at about US$25–30 million a year (Brown and Henry 1993).

But economists also recognize a range of other values, such as the value of keeping options open (option value), value in regard to one's descendants (bequest value) and the value of simply knowing that something is there (existence value). By combining all these it might be possible to arrive at a total economic value (TEV), but the operative word will always be *economic* (Hatch 1995). Total *worth* is another matter.

14.4 COMPROMISE AND CONSERVATION

If ecotourism is to benefit conservation it will be mainly because it provides an incentive to protect habitats from more destructive forms of use. Exactly the same reasoning lies behind all attempts to turn wildlife use to conservation advantage but ecotourism introduces yet another element into the concept of sustainability: that is, the quality of the tourist experience. If tourism is to support wildlife financially, then there must be something on which visitors can spend money, but it must not be something that will spoil the quality of the experience that visitors come to find.

Barnes *et al.* (1992) put forward four caveats for sensitive management of wildlife tourism:

- it needs to be sensitive to the scale and type of tourism, and the effects upon local cultures
- income from tourism needs to filter down to the local people whose lands and interests are affected
- where tourism occurs in protected areas, the goals of PA management must be furthered by the economic gains of tourism
- in developing countries, wildlife tourism should be accessible to visitors from a wide range of economic status and not restricted to the rich or foreign (following Groom *et al.* 1991)

The wildlife spectaculars of east and southern Africa have been able to withstand the onslaught of package tours because the areas involved are relatively large and, for the most part, tourists are confined to vehicles and designated routes as they pass through wildlife habitats. Other wildlife attractions, less robust or more difficult to display, can be destroyed, both ecologically and in terms of visitor satisfaction, by tourist development. Numerous surveys have shown that visitors to natural areas prefer minimal development and are disappointed by visual intrusions, noise and crowding (Buckley and Pannell 1990). Australian national parks, by world standards, are healthy and well managed and yet there is extensive evidence of visitor pressure (Table 14.1).

Table 14.1 Environmental impacts identified in Australian national parks (source: Buckley and Pannell 1990)

Disturbance to wildlife	Boat damage to waterway banks	Camp sites	Visual impacts: roads and buildings
Firewood collection	Water pollution	Human wastes	Weeds/fungi introduction
Noise	Changed water course	Litter	'Cultural vandalism'
Tracks and off-the-road vehicles	Water depletion	Trampling (human or horse)	Damage to archaeological sites

Perhaps, in small numbers, tourists of the type described by Ziffer would not spoil, for each other, the quality of the wilderness experience and would not do much ecological damage. But the attractions of a site have to be exceptional if the site is to generate money and local employment from small numbers of wealthy tourists. And who is to say that more money and employment opportunities could not be created by having more people, with more things to spend money on?

For every ecotourist destination there will be a position on the ecodevelopment playing field which, for conservation purposes, represents the best compromise with regard to site development, numbers of visitors, ecological impact and quality of visitor experience (Figure 14.1). The problem is not only how to determine this position, but how to maintain it against the inevitable pressures to grow.

14.5   IMPACT REDUCTION STRATEGIES

Butler (1991) has considered four possible strategies that site managers might adopt to prevent site degradation:

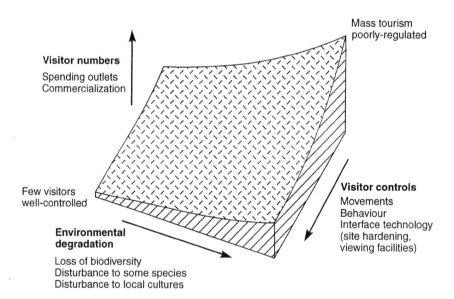

Figure 14.1 The 'ecotourism playing field': relationships between conservation, visitor numbers and site development can largely determine the quality of visitor experience.

- restrict or reduce tourist numbers
- select tourist types (low impact, high spending)
- educate tourists, hosts, entrepreneurs and government agencies to achieve impact minimization
- harden the resource

14.5.1 Tourist numbers

Some protected area (PA) authorities do impose a limit on visitor numbers but this is usually determined by the capacity of campsites and other accommodation within the PA. Visitation limits are less often based on other criteria. As Butler (1991) says: 'carrying capacities have been pursued almost as arduously as the Holy Grail, with about as much success'. The basic problem is that the people involved with tourism and PAs have different priorities. Those most concerned with generating revenue or local employment may not consider that a place is too crowded until it starts to lose popularity as a result of visitor numbers. Surveys of visitor satisfaction could remain encouraging for a long time because of 'recreational succession'. Year by year, visitors more tolerant of crowding replace those who are less tolerant so that repeated surveys continue to give acceptable results. This is one of the difficulties encountered in trying to determine the limits of acceptable change.

Conservationists are likely to favour a lower visitation pressure than those with a direct commercial interest, but they may still disagree about the limits, and they may not, in any case, carry much weight with the administrative authorities. Within the plan for a PA, how much visitor impact is acceptable? How important is it when ecological damage is so slight that only scientists can detect it? Where the main attraction is a waterfall, how much does it matter that a rare bird is getting rarer? And who is to decide?

14.5.2 Tourist selectivity

This runs contrary to the notion of having wildlife tourism accessible to people from all walks of life. The only practical way to discriminate in favour of small numbers of profitable tourists, who will have the least impact on the environment, is through pricing and the provision of minimal-impact facilities. In effect, this is an attempt to hold tourism development perpetually in phase two, as described earlier. Disregarding, for the moment, the question of whether this is a wise policy, the next question is whether the site can attract enough wealthy tourists to meet the financial targets. Unlike some exclusive venues, ecotourism sites do not attract a regular clientele. Ecotourists like to visit a different place each time.

Bhutan, which has opened up to international tourism relatively recently, has taken the course of only admitting small numbers of wealthier tourists to the country as a whole. This is being achieved through a minimum daily spending requirement which in 1992, at US$200, was more than six times the amount spent voluntarily by the average visitor to neighbouring Nepal (Wells 1993a).

Assuming that a site can attract enough visitors of the preferred type, there may be other reasons why a policy of limited development will be difficult to maintain:

(a) Local employment

Limited development is likely to provide limited employment opportunities for locals. Local support for conservation could decline as a result.

(b) Local economy

In remote regions it is commonly the case that local suppliers cannot dependably provide the quality goods and services which wealthier tourists expect. Under these circumstances, materials and provisions will be bought from the nearest urban centre and only a tiny proportion of tourism revenues is likely to reach the local economy. Alternatively, where a determined effort is made to 'buy local', the result can be inflated prices or commodity shortages for the local residents (Groom *et al.* 1991; Wells 1993b).

(c) Local perceptions

Although the support of local residents is particularly important for protected areas, the interest and support of the wider public also needs to be fostered. In many developing countries, as mentioned in section 14.3.2, PA admission charges for nationals are often a fraction of those which foreigners have to pay, otherwise few nationals would be able to afford to visit.

14.5.3 Education and training

Conservation education is extremely important and should help to ensure better management for conservation purposes. However, educating all those involved in developing and running an ecotourism site is a long-term strategy and, in the end, people will still make value judgements according to personal involvement. Those who have substantial investments to recover from a site are unlikely to see things in quite the same way as would a biologist with a special interest in invertebrates. It will always matter who makes the decisions.

In the shorter term, training of those who interact directly with the tourists could result in some marked improvements. There is a need for trained guides and interpreters who can set examples of behaviour, as well as being able to communicate with visitors. It is all too often the local guides, untrained and poorly paid, who disturb wildlife in order to please visitors by giving them a better view. Similarly, as Shackley (1994) observed, it is often the local people who strew litter because they have not yet become litter-conscious.

14.5.4 Site hardening and visitor control

Buckley and Pannell (1990) make the point that site hardening tends to accelerate recreational succession. Nevertheless, the most immediate measures that can be taken to reduce visitor pressure will include physical hardening of a site through construction of tracks, boardwalks, viewing platforms and barriers of various kinds. In some sites vehicles are relatively easy to control by low rock or log barriers and it is a fact that the majority of people are reluctant to carry picnic hampers very far from their vehicles. In other sites, off-the-road vehicles can be extremely difficult or expensive to contain. The sort of barriers needed to stop cross-country motor cycles would only be appropriate in exceptional circumstances.

A combination of education, appropriate regulation and surveillance, and good zonation planning can do a lot to direct visitor pressure away from the most ecologically sensitive areas. However, where this is done at the expense of visitor satisfaction, the social and psychological carrying capacities may be reached that much sooner.

Depending on the size and heterogeneity of an area, four or five different zones are usually enough to provide for:

- nature protection and research
- quiet, low-impact forms of tourism
- the more active or adventurous tourism

Facilities, visitor centres and other 'honeypots' must be arranged so that they draw visitors away from the quiet zones. Legal control over activities can be established through by-laws specific to the protected area.

The Federation of Nature and National Parks of Europe (FNNPE 1993) has described five zones considered to be appropriate for European PAs (Table 14.2) but the principles are the same everywhere. Zone 5, for example, is the equivalent of the buffer zone around some African PAs, where game cropping or safari hunting would be appropriate.

Zonation in time, as well as space, must sometimes be used to separate incompatible activities (e.g. angling and water sports) or for seasonal protection of wildlife. In the German–Luxembourg Nature Park, for instance, canoeing and rafting were destroying the habitat of the freshwater pearl mussel (*Margaritifera margaritifera*) by scraping the river bed at low water. These activities are now banned when the river depth falls below a critical level (FNNPE 1993).

The theory of zonation is largely a matter of common sense but, again, there is ample room for disagreement between people with different

Table 14.2 Proposed zones for sustainable tourism in and around Europe's protected areas (source: FNNPE 1993)

1 Sanctuary zone	2 Quiet zone	3 Compatible tourism. No extra development	4 Development of sustainable tourism	5 Zone outside PA where sustainable tourism is encouraged
Strictly protected from any form of tourism development	Access limited to small, mainly guided groups and few facilities provided	Existing activities and facilities continue if compatible with type of PA. No new development of facilities	Activities and facilities compatible with PA are developed. Developments all small scale and in keeping with local culture and building styles	Immediately outside the PA, developments similar to zone 4 are encouraged to maintain quality of PA

management priorities. Moreover, as Butler (1991) observed, once development of an area has progressed, it is rarely possible to turn the clock back and start again. In Antarctica, those involved in ecotourism are trying from the start to get things right. The circumstances are exceptional but there are some useful lessons with wider application.

14.6 ECOTOURISM IN ANTARCTICA

The first genuine Antarctic tourists flew from Punta Arenas in Chile in 1956. By the early 1990s there were about 10 000 visitors a year being landed from ships and there could be double or treble that number by the end of the century (Stonehouse 1994).

Tourism by boat generally conforms to what has come to be known as the Lars-Eric Lindblad pattern. Ships with a capacity of 100–140 passengers are used and tourists are taken ashore in groups of 10–15 using inflatable dinghies. Each group is then accompanied by a tour guide, and although individuals are allowed to wander away from the group to approach animals, take photographs, or simply be apart, they are expected to comply with a strict code of conduct. According to Stonehouse (1994) the procedure has worked very well but as the number of ships increases it may not be possible for operators to keep finding quiet spots to put their groups ashore. Perhaps it will be necessary to develop special sites where larger numbers of tourists can be put ashore together.

The International Association of Antarctic Tour Operators (IAATO) was formed in 1991 and seeks to coordinate itineraries and shore visits so as to minimize interference between members. The Association also publishes Guidelines of Conduct for passengers and tour operators (Splettstoesser and Folks 1994). Extracts from the Guidelines with relevance for wildlife appear in Box 14.1. Although Antarctica is unique, the example of cooperation in pursuit of sustainable standards could be followed to some extent at national and regional levels everywhere.

14.7 LOCAL PEOPLE AND ECOTOURISM

There is general agreement in the literature that local residents should have a fully participatory role in every stage of development of an ecotourism (or other ecodevelopment) proposal. The wisdom of this recommendation is not being disputed here but neither should it be thought that local participation is a proven solution to all problems. The move towards full participation has arisen largely because of increasing difficulty and failure with other approaches. A new track record has yet to be established.

In 1992 the World Bank published what amounted to a progress report on integrated conservation–rural development projects (Wells *et al.* 1992). Twenty-three projects from Asia, Africa and Latin America were selected as case studies. All had development components linked to PAs; all had been in operation for at least three years and all were considered to be among the most promising and effective ecodevelopment projects of their type. Even so, it was not possible to demonstrate that development initiatives had anywhere resulted in diminished pressure on protected areas. Evidently it is not enough to try to raise living standards in the hope that wildlife and wild places will somehow benefit; conservation objectives must be clearly specified, targeted and monitored (Bolton 1994). Ecotourism projects, even with full local participation, must still target the crucial linkage between conservation and rural development if the intention is to use tourism for the benefit of conservation.

The Annapurna Conservation Area Project (ACAP) in Nepal has gained a deserved reputation as an innovative attempt to integrate conservation and ecotourism with rural development. The ACAP was legally gazetted in 1992 and an NGO, the King Mahendra Trust for Nature Conservation (KMTNC), became the official management authority for a 10-year period. The concept of a multiple use 'conservation area' is acceptable to a majority of the 40 000 residents because it is seen as emphasizing the role of villagers in using and managing natural areas. National parks, in contrast, have become associated with the imposition of restrictions. In view of the importance of public perceptions, this attention to terminology was very

Box 14.1   Some Antarctica tour operator guidelines

- Enforce the visitor guidelines in a consistent manner, bearing in mind that guidelines must be adapted to individual circumstances. For example, fur seals with pups may be more aggressive than without pups so visitors should stay further away
- Hire a professional team. Place an emphasis on lecturers and naturalists who will not only talk about the wildlife, history and geology, but also guide passengers when ashore
- Ensure that for at least every 20–25 passengers there is one qualified naturalist lecturer/guide to conduct and supervise small groups ashore
- Limit the number of passengers ashore to 100 at any one place and time
- It is the responsibility of the tour operator to ensure that no evidence of visits remains behind. This includes garbage of any kind
- Respect historic huts, scientific markers and monitoring devices

Some Antarctica Visitor Guidelines

- Do not disturb, harass, or interfere with the wildlife
  - Never touch the animals
  - Maintain a distance of at least 15 ft (4.5 m) from penguins, all nesting birds and true seals, and 50 ft (15 m) from fur seals
  - Give animals right of way
  - Do not position yourself between a marine animal and its path to the water, nor between a parent and its young
  - Stay outside the periphery of bird rookeries and seal colonies
  - Keep noise to a minimum
  - Do not feed the animals, either ashore or from the ship
- Do not walk on or otherwise damage the fragile plants, i.e. lichens, mosses and grasses. Damage from human activity among the moss beds can last for decades
- Leave nothing behind and take only memories and photographs
  - Leave no litter ashore
  - Do not take souvenirs, including whale and seal bones, rocks, fossils, plants, other organic material, or anything that may be of historical or scientific value
- Do not interfere with protected areas or scientific research
- Do not smoke during shore excursions (fire hazard)
- Stay with your group or with one of the ship's leaders when ashore

Source: Extracted and abbreviated from Splettstoesser and Folks (1994)

wise, although the area is so large (4600 km^2) and diverse that designation as a national park would not have precluded management for 'intensive use', or any of the other management objectives for which zones have actually been declared.

So far the KMTNC has made an admirable start under very difficult circumstances and with a limited budget, but it is too early to say whether the conservation of biological diversity, or even the conservation of the more conspicuous wildlife species, will be well served by the ACAP. Project activities, such as replacing wood stoves with kerosene stoves, establishing forest management committees, encouraging stall feeding of livestock and the planting of fodder trees, could help to conserve biological diversity to the extent that the activities are favourable to that outcome. However, biodiversity conservation was not specifically included in the original objectives of the ACAP and any benefits will be difficult to quantify in the absence of biological monitoring (Wells 1993b). Furthermore, 'basing actions on local decisions has meant that the process of developing new conservation regulations and resource use practices has been very localized and slow' (Stevens and Sherpa 1992 cited in Wells 1993b).

14.8 TOURISTS AND THE WILDLIFE RESPONSE

It can be difficult to predict how a species will respond to continued and increasing human presence. Some animals, if they can, may simply move out of the area; others may become more arboreal or nocturnal, while others may show little concern or even become habituated (Griffiths and Van Schaik 1993). It is even more difficult to predict long-term ecological changes that may result from these kinds of responses so it makes sense to minimize disturbance.

14.8.1 Birds

The effects of human presence on bird nesting colonies has been extensively studied in temperate regions. Effects have included nest desertion, decreased hatching success and increased risk of predation. Where human disturbance has been varied experimentally, fledging success has been inversely related to amount of disturbance (Burger and Gochfeld 1993).

Colonies of nesting seabirds have been found to be able to tolerate large numbers of visitors. In Australia, Michaelmas Cay, off the north Queensland coast, is an island of only 1.8 ha and yet some 30 000 birds of seven species can be nesting there at the peak nesting season. The birds nest on a vegetated area of 300 × 60 m which is fenced off from the beach.

The cay and surrounding coral reef is one of the main bird nesting islands in the Great Barrier Reef Marine Park. The management plan stipulates that there may not be more than 100 visitors on the beach at any one time but up to 150 have been recorded. The tiny anchorage is frequently crowded and the beach and reef at times have the colour and atmosphere of a fun resort. The impact of this disturbance on the different bird species is unknown but the colony, as a whole, shows no obvious decline. However, public responses indicate widespread dissatisfaction with the crowding and a need for better interpretation of park values (Muir and Chester 1993).

In the Galapagos islands, Burger and Gochfeld (1993) studied three species of booby (*Sula dactylagra*; *S. Sula*; *S. nebouxii*) which were considered to be so tame as to be untroubled by tourists. However, there were significantly fewer nests within 10 m of footpaths and, depending on the species, birds which were displaying within 2 m of paths flew or walked from their positions 62–95% of the times that visitors passed. The effect, if any, on reproductive success is not known.

14.8.2 Mammals

Tours to view marine mammals are very much under the control of the boat operators so it should be relatively easy to limit disturbance. Beluga whales (*Delphinapterus leucas*) in the St. Lawrence River estuary, Canada, respond to ecotourism boats by spending longer intervals submerged, bunching into groups and increasing their swimming speed. The avoidance behaviour intensifies with increasing number of boats and boat speed (Blane and Jaakson 1994).

Forest mammals are notoriously difficult to display to visitors. In Nepal, tigers in the Chitwan National Park were attracted, after dark, to buffalo baits. To see a tiger at the bait, visitors to the Tiger Tops Jungle Lodge would be led, barefoot, in small, quiet groups, along a swept footpath to a lookout. From there they would be shown the tiger by spotlight. In Kanha National Park, India, natural tiger kills are located by trackers in the early morning and visitors are taken on elephant back to see the tiger sleeping off the meal. Some tigers have become well accustomed to being surrounded by elephant loads of tourists and no negative effects have been reported. Baiting of lions for tourists in India's Gir forest, however, may have exacerbated the lion–human conflicts there (Saberwal *et al.* 1994).

Animals which occur in higher densities, such as deer and forest antelopes, can be seen relatively easily at waterholes, and will not generally be aware of visitors if lookout hides are raised a few metres above ground level. This arrangement will not succeed if people are constantly coming and going from the hides but it works well if groups enter and leave together and keep reasonably quiet while they are in place. In Africa this method is used to display shy forest animals such as the African Nyala (*Tragelaphus angasi*), bongo (*Boocercus euryceros*) and giant forest hog (*Hylochoerus meinertzhageni*).

Noisy, arboreal animals are not difficult to locate and can be viewed, like more conspecuous Madagascar's lemurs, by simply staring up at them.

Kinnaird and O'Brien (1996) describe the responses of black macaques (*Macaca nigra*) and spectacled tarsiers (*Tarsier spectrum*) in Tangkoko DuaSudara Nature Reserve in North Sulawesi, Indonesia. Three monkey troupes, all habituated to humans, were studied and it was found that each troupe had a different level of tolerance for groups of tourists. The biggest troupe (97 individuals) tended to split into sub-groups and move away if tourists came in groups of more than seven people. The smallest troupe (49) had less tolerance of people in groups of more than seven, and the third monkey troupe (60) invariably fled from tourist groups of more than five people. Thirty per cent of the monkeys in this last troupe would climb trees in response to people who came only in twos or threes.

Tarsiers, being nocturnal, were viewed by torchlight as they emerged from their tree nests at dusk. Visitors with flashlights and photoflashes gathered in numbers of up to 30 around the nesting trees and possibly delayed the tarsiers' emergence. Tarsiers temporarily abandoned one sleeping tree after large groups of tourists had visited on successive nights.

Some degree of tameness will probably make animals easier to see but it can get out of control. Habituated primates around tourist camp sites can take to snatching food, raiding tents and generally being more than a nuisance (tame ostriches at one African site had to be got rid of for this reason). Habituation can also affect the animals' ecology if it disrupts normal foraging behaviour and animals are encouraged to wait for titbits from visitors. The habituation of gorilla groups in Rwanda is generally held to be an exemplary way of doing things. Small, well-briefed groups of tourists are taken by experienced guides to see the gorillas. Even this procedure, however, involves the risk of infecting the gorillas with diseases carried by humans (Woodford, Chapter 12, this volume).

14.9 CULTURE

To the extent that cultural concerns create positive or negative attitudes towards a PA, wildlife conservation could be affected. Nevertheless, biological values and cultural purity are not the same thing and the commercialization of local traditions is not necessarily inimical to conservation.

Local cultures can be surprisingly resilient to western influence. The Sherpas of Nepal, for instance, despite their involvement with more than 11 000 trekking tourists and 40 mountaineering expeditions a year, are said to have retained their social and cultural cohesion remarkably well (Wells 1993a). The people of the Annapurna region of Nepal are reported to have shown less cultural resistance to western influence (Gurung and De Coursey 1994).

Myra Shackley (1994) examines the interrelationships between tourists and local people in the newly opened Mustang region, to the north-west of

Annapurna, and her findings are consistent with general observations in other developing countries. The relatively remote villages in the east of the region had received less than a tenth of the tourism which the western region had experienced. The people were friendly and hospitable and there was a reciprocal cultural curiosity between visitors and locals. In the more accessible western parts, some negative consequences of tourism had begun to show. On the part of the visitors these included:

- improper dress and behaviour offensive to local customs
- ignorance of required behaviour at sacred sites
- intrusive or thoughtless photography, often without asking permission
- casual intrusion into homes
- encouraging begging by offering sweets or money to children

Unwelcome responses on the part of the local people included:

- begging by children for food, money, sweets or pens
- pestering visitors to buy souvenirs
- 'mobbing' visitors who try to take photos without paying
- refusal to cooperate for photographs, even when money was offered
- use of sexual swearwords and gestures learned from visitors

Shackley (1994) suggests that ethnic tourism carries the seeds of its own destruction and that the critical threshold between agreeable reciprocal interest and mutual exploitation may be as low as 50 visitors for some communities.

'Mutual exploitation', provided it takes an acceptable form to the parties involved, is not necessarily detrimental to the success of a PA and the conservation of wildlife. On the contrary, commercialism may provide the best incentive for preserving cultural traditions as well as for protecting wildlife. This appears to be true with regard to the ceremonial dresses, dances and the production of artefacts by some communities in Papua New Guinea. Examples from other parts of the world can be found in MacCannell (1984). It is easy to be cynical about commercial pseudo-culture but the alternative might be to lose traditions altogether.

A more unfavourable scenario for conservation would be the exploitation of a local culture when the locals have no wish to reciprocate. If locals resent the intrusion and attention of outsiders, they might also resent the existence of a protected area which attracts visitors. In that case the cultural dimension would be an obstacle to wildlife conservation. This situation is most easily avoidable where people and wildlife are well separated. Park authorities in the Manu Park of Peru have discouraged contact between tourists and indigenous communities for fear of cultural pollution and the introduction of diseases to isolated groups (Groom *et al.* 1991). Cultural concerns have also been expressed on behalf of Australian Aborigines (Altman and Finlayson 1993; Hall 1994).

14.10 CONCLUSIONS

Ecotourism is no more a panacea for conservation problems than are other forms of wildlife utilization but it has already proved its worth in many PAs and could help to bring some other sites under protection. It is not appropriate at every site and, where it is appropriate, the detailed recipe for success will be site-specific.

Because of the different and often conflicting interests of the numerous stakeholders in tourism projects, there needs to be some mechanism by which the interests of the different stakeholders can be represented in planning and policy formulation. Even then, ecotourism can be more difficult to regulate than some other forms of wildlife utilization, and the potential for conservation can be difficult to realize. Indeed, if attendance at workshops and conferences on ecotourism is any indication, then biological interests are not particularly well represented in the industry. And if commercial stakeholders are untroubled by questions about ecological sustainability, it is, perhaps, because they are confident in regarding ecotourism as a non-consumptive use of wildlife.

Conservation objectives have rarely, if ever, been specifically targeted in ecotourism, or other ecodevelopment projects, so neither has there been any monitoring of progress towards conservation goals. There is a need for the linkage between local development and conservation to be established for ecotourism projects as well as for other wildlife utilization schemes.

In and around PAs, because of the failings of top-down approaches to conservation as well as for moral and ethical reasons, there is a growing demand for bottom-up solutions to land-use conflicts so that local residents become fully involved as project participants and beneficiaries. Unfortunately, local participation is no guarantee of swift or certain success in conserving wildlife or biodiversity; and ecotourism, with or without local participation, cannot always produce the desired results in terms of local perceived net benefits. At best, ecotourism, as an ecodevelopment measure, will demonstrate the local benefits of maintaining wildlife and natural habits for the long term. The costs of protection, as they are seen locally, will not then be resented. One of the dangers, as mentioned in section 4.2.5, is that of overselling a project and creating unrealisitc expectations. This could lead to a PA being dismissed, in local minds, as a failure.

REFERENCES

Altman, J. and Finlayson, J. (1993) Aborigines, tourism and sustainable development. *Journal of Tourism Studies*, **4** (1), 38–50.

Barnes, J., Burgess, J. and Pearce, D. (1992) Wildlife tourism, in *Economics for the Wilds* (eds T.M. Swanson and E.B. Barbier). Earthscan Publications, London, pp. 136–51.

Blane, J.M. and Jaakson, R. (1994) The impact of ecotourism boats on the St. Lawrence beluga whales. *Environmental Conservation*, 21 (3), 267–9.

Bolton, M. (1994) Ecodevelopment in the crowded tropics: what prospects for conservation? *Environmental Conservation*, 21 (3), 259–62.

Brown, G. and Henry, W. (1993) The viewing value of elephants, in *Economics and Ecology: New Frontiers and Sustainable Development* (ed. E.B. Barbier). Chapman & Hall, London, pp. 146–55.

Buckley, R. and Pannell, J. (1990) Environmental impacts of tourism and recreation in national parks and conservation reserves. *Journal of Tourism Studies*, 1 (1), 24–32.

Burger, J. and Gochfeld, M. (1993) Tourism and short-term behaviour responses of nesting masked, red-footed and blue-footed boobies in the Galapagos. *Environmental Conservation*, 20 (3), 255–9.

Butler, R.W. (1991) Tourism, environment and sustainable development. *Environmental Conservation*, 18 (3), 201–9.

Cater, E. (1994) Introduction, in *Ecotourism: A Sustainable Option?* (eds E. Cater and G. Lowman). Wiley & Sons, Chichester, pp. 3–17.

Durst, P.B. (1994) Planning for ecotourism within the framework of the Tropical Forests Action Programme. *Tiger Paper*, 11 (2), 7–14.

FNNPE (1993) *Loving Them to Death? Sustainable Tourism in Europe's Nature and National Parks*. Federation of Nature and National Parks of Europe, Grafenau, Germany.

Giannecchini, J. (1993) Ecotourism: new partners, new relationships. *Conservation Biology*, 7 (2), 429–32.

Griffiths, M. and Van Schaik, C.P. (1993) The impact of human traffic on the abundance and activity periods of Sumatran rain forest wildlife. *Conservation Biology*, 7 (3), 623–6.

Groom, M.J., Podolsky, R.D. and Munn, C.A. (1991) Tourism as a sustained use of wildlife: a case study of Madre de Dios, Southeastern Peru, in *Neotropical Wildlife Use and Conservation* (eds J.G. Robinson and K.H. Redford). University of Chicago Press, Chicago, pp. 393–412.

Gurung, C.P. and De Coursey, M. (1994) The Annapurna Conservation Area Project: a pioneering example of ecotourism? in *Ecotourism: A Sustainable Option?* (eds E. Cater and G. Lowman). Wiley & Sons, Chichester, pp. 177–94.

Hall, C.M. (1994) Ecotourism in Australia, New Zealand and the South Pacific: appropriate tourism or a new form of ecological imperialism? in *Ecotourism: A Sustainable Option?* (eds E. Cater and G. Lowman). Wiley & Sons, Chichester, pp. 137–57.

Hatch, J. (1995) Economic valuation and analysis of wildlife use, in *Conservation Through Sustainable Use of Wildlife* (eds G.C. Grigg, P.T. Hale and D. Lunney). Centre for Conservation Biology, The University of Queensland, pp. 99–105.

Hawkins, D. (1991) Address to Ecotourism Management Workshop, held at George Washington University, June 1991, cited in Giannecchini (1993).

Kinnaird, M.F. and O'Brien, T.G. (1996) Ecotourism in the Tangkoko DuaSudara Nature Reserve: opening Pandora's box? *Oryx*, 30 (1), 65–73.

MacCannell, D. (1984) Reconstructed ethnicity: tourism and cultural identity in Third World Communities. *Annals of Tourism Research*, 11, 375–91.

Muir, F. and Chester, G. (1993) Managing tourism to a seabird nesting island. *Tourism Management*, **14** (2), 99–105.

Navrud, S. and Mungatana, E.D. (1994) Environmental valuation in developing countries: the recreational value of wildlife viewing. *Ecological Economics*, **11**, 135–51.

OECD (1994) *Tourism Policy and International Tourism in OECD Countries 1991–1992*. OECD, Paris.

Prosser, R. (1994) Societal change and the growth in alternative tourism, in *Ecotourism: A Sustainable Option?* (eds E. Cater and G. Lowman). Wiley & Sons, Chichester, pp. 19–37.

Saberwal, V.K., Gibbs, J.P., Chellam, R. and Johnsingh, A.J.T. (1994) Lion–human conflict in the Gir Forest, India. *Conservation Biology*, **8** (2), 501–7.

Shackley, M. (1994) The Land of Lo, Nepal/Tibet: the first eight months of tourism. *Tourism Management*, **15** (1), 17–26.

Splettstoesser, J. and Folks, M.C. (1994) Environmental guidelines for Antarctica. *Annals of Tourism Research*, **21** (2), 231–44.

Stevens, S.F. and Sherpa, M.N. (1992) Indigenous peoples and protected area management: new approaches to protected area management in Nepal. Paper presented at the World Parks Congress, Caracas, 10–21 February 1992.

Stonehouse, B. (1994) Ecotourism in Antarctica, in *Ecotourism: A Sustainable Option?* (eds E. Cater and G. Lowman). Wiley & Sons, Chichester, pp. 195–212.

Wells, M. (1992) Biodiversity conservation, affluence and poverty: mismatched costs and benefits and efforts to remedy them. *Ambio*, **21** (3), 237–43.

Wells, M. (1993a) Neglect of biological riches: the economics of nature tourism in Nepal. *Biodiveristy and Conservation*, **2**, 445–64.

Wells, M. (1993b) A profile and interim assessment of the Annapurna Conservation Area Project, Nepal. Paper prepared for the Liz Claiborne Art Ortenberg Foundation Community-Based Conservation Workshop. Airlie, Virginia.

Wells, M., Brandon, K. and Hannah, L. (1992) *People and Parks: Linking Protected Area Management with Local Communities*. World Bank, Washington, DC.

WTO (1992) *Tourism Trends and the Year 2000 and Beyond*. Presentation for the World Trade Centre, Seville.

Ziffer, K.A. (1989) *Ecotourism: The Uneasy Alliance*. Working Paper No. 1. Conservation International, Washington, DC.

Part Three

A Synthesis

-15

Synthesis and conclusions

The question of whether using wildlife is good or bad for conservation can never be answered with an unqualified yes or no. It is not a yes/no sort of question. But to oppose all commercial use of wildlife is to close off options for conservation where and when the protected area strategy fails. It will be concluded here that all options must be kept open and we must develop a battery of questions by which the pros and cons of any particular case can be assessed. In other words, we need to be able to recognize and predict the circumstances under which utilization can be of long-term benefit to conservation. There are a great many factors to be taken into account. Luxmore and Swanson (1992) ask the 'critical analytical question': 'What are the characteristics of human intervention that render it a force for conservation rather than a force for conversion?' But that question does not go far enough because the human interventions that might succeed with one species can be hopelessly wrong for another. Management strategies only make sense in relation to the characteristics of the targeted species.

There is not much to be compared between the ranching of butterflies and the harvesting of whales but both are examples of wildlife being used for commercial purposes. Many people who would oppose whaling would no doubt approve of butterfly ranching and it is most unlikely that those who declare themselves opposed to all commercial use of wildlife would object equally strongly to the two cases. It may be instructive, therefore, to consider how the two cases differ. Table 15.1 sets out some of the contrasts.

The contrasted features of butterfly ranching and whale harvesting are only a few of the considerations which could be used in assessing conservation potential. It may bring some order to what would otherwise be a random checklist if the concerns are grouped under the headings which have emerged in earlier chapters.

Conservation and the Use of Wildlife Resources. Edited by M. Bolton.
Published in 1997 by Chapman & Hall. ISBN 0 412 71350 0.

Table 15.1 Some contrasted parameters of whale harvesting and butterfly ranching

Butterfly ranching	*Whale harvesting*
Minor habitat manipulations to favour target populations. No obvious detriment to other species	No habitat interventions possible. Harvesting entirely dependent on natural productivity
Species *r*-selected. Rapid population recovery possible	Species *K*-selected. Very slow population growth
Harvest of sub-adult life stage	Harvest of adults
Local populations of species relatively easy to monitor	Populations difficult to monitor
Cottage industry, small operations. Mainly in developing countries	Corporate industry, high investments. Mainly by industrialized countries
Often on private or communal land	Largely in high seas 'commons'
Insect welfare not a public concern	Humane killing difficult to achieve. Whales high public-profile sentient animals
Scarcely any publicity	Much publicized

15.1 CONSERVATION BENEFIT

Benefit, in the sense of gain, must be assessed relative to some sort of baseline which represents the alternative, or expected alternative, to the utilization which is being considered. If a habitat under threat can be saved by declaring it a Protected Area, there may be no need to consider any other form of use. Otherwise, if a utilization scheme is the option, the benefits need to be considered relative to the threatened alternative. That is how the benefits of a PA would be assessed.

Benefits vary in kind as well as degree and can be achieved indirectly as a result of financial gain. The benefits of less intensive forms of use could include the following :

- serving to increase the perceived value of natural habitats
- helping to retain the integrity of natural habitats and preventing their conversion
- encouraging the creation and restoration of habitats for wildlife
- encouraging private investment (money and time) in conservation and so reducing cost to the public purse
- contributing to local, regional and national economies

- increasing the incentives (through increased numbers of stakeholders and lobby groups) for governments to favour conservation policies
- economical control of pest species
- encouraging research and development in wildlife management

Some of the listed benefits might be somewhat double-edged and no doubt a list of risks could also be compiled, but the risks also would need to be assessed relative to realistic alternatives. Rarely are commercial interests totally aligned with those of biodiversity conservation. A commercial fishery is considered successful if it is sustainable and believed to be doing less harm than would the management alternatives. Outside PAs it is unrealistic to expect fish to be left undisturbed.

Unquestionably, the modification of intensive agriculture to cater for field sports can have widespread benefits for conservation but only in areas suitable for game and with management priorities favouring the target species. Nevertheless, in terms of broad ecological benefits, management for field sports offers the most widespread example of conservation through utilization. The case study of partridges in England demonstrates particularly well a direct link between wildlife use and conservation (Aebischer, Chapter 8, this volume). Similar benefits might possibly be obtained in some developing countries through the live reptile and bird trade but the appropriate management measures will need to be identified and demonstrated to those involved.

The more intensive forms of use, with an emphasis on specialization and productivity, may be less beneficial to the broader ecology of an area, but benefits could include:

- greatly increasing the numbers of the target species
- helping to reduce pressure on other populations of the target species
- promoting detailed research into the target species' requirements

Again, there are counter-arguments and it could be said that increasing the commercial value of a species is likely to increase the pressure upon it. It is at this level that all the details of a proposal must be taken into account. There is no reason to think that the ranching of crocodiles, for instance, has been detrimental to their overall security in the wild. The greater risk of specialized production may well be that it will encourage selective breeding and domestication instead of building a commitment to foster the wild product. This is already happening with a few species, notably the African grasscutter (*Thryonomys swinderianus*) (Jori *et al.* 1995). Another edible rodent, the capybara (*Hydrochaeris hydrochaeris*), may not be far behind on the road to domestication (Ojasti 1991).

Captive production, depending on the form it takes, may have very limited conservation value. True, a healthy captive population can serve to:

- act as a source of specimens for reintroductions

- facilitate research and stimulate interest in the species
- provide excellent opportunities for education

But even these benefits are eroded by selective breeding and domestication. It was once flippantly remarked to the writer that the best thing that ever happened for turkeys was Christmas. It is a point well made but the festive season only guarantees large numbers of domestic birds. It is doubtful whether domestication has improved security for the wild counterparts of any existing domestic animal. Not that domestic animals are unable to revert to the wild: some have retained a remarkable genetic potential in this respect. It was fascinating to see, in a tropical Australian garden, a domestic hen communicating with her scattered chicks while a large bird of prey was overhead. Apparently in response to soft croaking calls from the hen, the chicks promptly 'froze' and remained motionless instead of running as they would have done in response to other disturbances. The birds were Old English Game bantams and one can only guess how many generations of the breed have passed since this behaviour had any significance in terms of genetic selection.

15.2 STATUS OF TARGET SPECIES: PESTS OR RESOURCES?

Section 4.1.5 noted the difficulties of integrating recreational hunting with pest control. The same sort of conflict can exist in attempting to integrate the control of a pest with its commercial exploitation, but for commercial purposes it is easier to quantify the problems and opportunities involved. With field sports some of the parameters (quality of experience, hunter satisfaction) are relatively intangible.

By definition, a pest species must have the potential to cause a problem; pest control operations are intended to ameliorate that problem. Killing pests is therefore a means to an end, not an end in itself. It is rarely feasible to exterminate a pest completely because getting rid of the last few can be inordinately expensive. Accepting that eradication is usually impracticable, it follows that there will generally be a need for continued control operations. For pest species that are saleable this invites the question as to whether a sustained commercial harvest would also satisfy, or complement, the pest control objectives. In short, when can a pest become a useful resource?

Allen *et al.* (1995) examine the opportunities presented by feral goats, feral pigs and rabbits in Queensland and conclude that pest species are more suited to commercial harvesting when:

- the resources and expertise required for capture, handling and transport already exist in conventional livestock industries
- customary methods of control do not conflict with commercial harvesting, and harvesting practices neither enhance nor prevent effective control

- the landowner/manager derives monetary gain from the harvested animals. The 'pest' must make a positive contribution to the economy of the managed land

In the case of rabbits it was concluded that commercial harvesting was not compatible with control. Harvesting methods (shooting and trapping) are ineffective in regulating rabbit numbers, and control by disease (myxoma and calicivirus) is incompatible with commercial harvesting.

In some species harvesting may become unprofitable before pest densities are reduced to the levels that would be achieved by conventional control methods. In such cases harvesting and control could be used in a complementary way, with the latter as a follow-up to the former, on an area-rotation basis. Alternatively, where pest control is carried out at public cost, it could be more cost-effective to subsidize commercial use of pests than to fund control operations. The subsidy would be calculated to extend commercial viability to lower pest densities (Choquenot *et al.* 1995).

The commercial use of pest species, compared with pest control, is likely to occur in agricultural land and may not yield conservation benefits because the ecological outcome will be much the same. The conservation levy raised from harvesting monkeys in Mauritius (Stanley and Griffiths, Chapter 11, this volume) is an advantage of utilization over pure pest control. In different situations other gains are possible. Grigg (1995) has conservation firmly in mind when he argues in favour of developing a kangaroo meat industry. If kangaroos and sheep together can be regarded as a managed resource, then total grazing pressure on the arid and semi-arid rangelands might be reduced without loss of income to the graziers. As things are, graziers must accept kangaroos as a liability and any reduction of sheep numbers in favour of kangaroos would be unthinkable.

It could also benefit conservation in and around PAs in African and Asian countries if crop-raiding animals were not being killed to go to waste. Attitudes towards the PAs might then be a little less negative and PA management might be easier and less costly as a result.

An improvement in the welfare of a pest animal could also result from commercial harvesting if it brought the species under veterinary and other regulations governing commercial handling and slaughter of animals (Woodford, Chapter 12, this volume). As Caughley and Sinclair (1994) have noted, 'the notion of humane treatment is often the first casualty of turning a species into a pest'.

15.3 SUSTAINABILITY

Because circumstances are always changing, management needs to be adaptive. Adaptive management depends on monitoring and the capacity to make responsive changes. In the case of field sports most management has been based on trial and error but overharvesting has generally been avoided because: traditional game species have a high intrinsic rate of increase, population growth is generally strongly density dependent, and recreational shooters and managers have been anxious not to reduce breeding populations (Caughley and Sinclair 1994). This combination of ecological and uniquely human factors serves as a reminder that as many as four broad components of sustainability can be distinguished and may need to be managed adaptively (section 3.5).

15.3.1   The ecological component

The biological characteristics of the target species are the fixed parameters in any utilization scheme based on wild animals. This can be significant for non-consumptive as well as consumptive use. Some animals obviously have far more ecotourism potential than others, and some are much easier to protect than others. The success of the 'Earth Sanctuaries' mentioned in section 13.5.6 is partly attributable to the fact that the animals are resident in quite small areas and their populations respond well to the exclusion of feral cats and foxes.

There may be some important biological questions that simply cannot be answered at the beginning of a project but it is easy enough to indicate the sort of questions that are relevant.

(a) Niche and life history

Taking a harvest of wild animals disturbs a web of natural interactions and it is not always possible to predict the ecological consequences. There will certainly be some compensatory changes, and the examples from fisheries (Chapter 3) emphasize the need to proceed cautiously and have a monitoring programme in place. On the other hand, the ecological consequences of harvesting may be quite subtle. Webb (1995) draws attention to the fact that when crocodiles were largely removed from the tidal waters of Northern Australia no dramatic changes were noticed over a period of 20 years. Whatever changes may have occurred they were insignificant compared with what happens when wetlands are converted for agriculture.

The life history of target species can have more obvious implications for harvesting; some segments of populations being more expendable than others. It takes millions of larvae to replace one giant clam which is safe from predation. Where gulls' eggs have been collected traditionally, it has long been known that the first eggs laid by a bird will usually be replaced so that the drain on the population is minimized. Mrosovsky (1989) refers to the thousands of turtle eggs that are destroyed on some beaches by other excavating turtles. He suggests that turtles and local people might benefit if a third of these doomed eggs were collected for food and the remainder were protected. Similarly, Thomsen and Brautigam (1991) mention that the hyacinth macaw (*Anodorhynchus hyacinthus*) lays two eggs but never succeeds in raising both young. In any proposed ranching scheme it would be worth monitoring the consequences of harvesting one egg from each nest.

(b) Intrinsic rate of increase

Ultimately, this sets the upper limit to the sustainable rate of harvest and it can be a serious constraint to sustainable use. For example, elephants, with an intrinsic rate of increase of 6–7% and a maximum sustainable yield of around 3–4%, have been described as terrestrial whales because more money could be made by converting them to cash and reinvesting the capital. The same would apply to rhinos (Caughley 1993).

(c) Density-dependence, sex and behaviour

The extent to which animal populations compensate for harvested losses by an accelerated rate of increase is obviously important, but the possibility that environmental factors could be overriding must always be borne in mind. The winter mortality of partridge depends on spacing behaviour in relation to nesting habitat but the annual survival of partridge chicks depends on the weather and the abundance of insects (Aebischer, Chapter 8, this volume). The weather can also be the main influence on some large mammal populations; the saiga antelope is not unique in this respect. Spinage and Matlhare (1992) warn that the game herds of the Kalahari should not be harvested without regard to the recurring drought-induced mortality which is unlikely to be density-dependent. In an Australian example, there is a poor relationship between rate of increase and density for red kangaroos (*Macropus rufus*) in Queensland. Available food for these large kangaroos in the arid and semi-arid rangelands correlates much more closely with rainfall than with density (Cairns and Grigg 1993). Contrast this with the density-dependent responses of red deer populations in the well watered Scottish Highlands. The red deer herds also indicate the importance of sex ratio in relation to management objectives (Reynolds and Staines, Chapter 10, this volume).

15.3.2 The socio-economic component

The requirements for ecological sustainability and socio-economic sustainability can be difficult to reconcile and this presents a fundamental problem. Farmers of all kinds understand the importance of having a strong market demand for their produce; they readily subscribe to market research and development. They understand the value of cooperative marketing and having some control over levels of supply – given that farmers have only limited control over total production. Commercial harvesters of wildlife have even less control over production; they must be constrained against over-exploitation and, to the extent that the objectives of biodiversity conservation are to be met, they should not try to produce more by becoming too specialized because that is the road to habitat conversion!

There will be no problem if a wildlife product is in great demand and short supply. All too often, however, wildlife harvesters are likely to be helpless pedlars, offering their produce for what they can get instead of being able to target (let alone influence) customer demands. This does not make for sustainability. Elliott and Woodford (1995) have described exactly this situation with reference to the harvesting of feral goats in Australia. More than a million feral goats were harvested in 1992–93 but the harvesters operate as an assemblage of hunter–gatherers; price takers rather than price makers, with no buffer against the vagaries of the market.

With feral goats in Australia there are opportunities for developing a more sophisticated industry which does not conflict with conservation objectives, but it is not always so. Where ecotourism, for instance, is to be developed with conservation as a major goal, the problem will be acute: how to remain sustainably product-based and able to resist becoming market-led.

It would be difficult to overestimate the importance of markets even in purely commercial terms. It was long ago recognized that in African savannah a variety of browsing and grazing ungulates would be more productive of meat than cattle could ever be, but few game cropping schemes have shown big enough profits to encourage further investment (Eltringham 1994). It was once profitable to farm ostriches primarily for their feathers but that changed with the whims of fashion. There will always be an element of risk and uncertainty under the heading of socio-economics. There may also be opposing forces to overcome in finding or creating a market niche. In Australia, those attempting to establish the human consumer market for kangaroo meat encounter a distinct lack of cooperation from the beef industry.

15.3.3 The community component

The importance of meeting the needs of local communities has been repeatedly stressed. However, lifestyles and traditional values at community level can change quite quickly – with direct consequences for local enterprises. The Taiwanese communities who were once keen to make money from butterflies have now moved on to other things (New, Chapter 6, this volume) but sometimes traditional ways can be disrupted by outsiders. The collection of eggs of megapodes (Megapodiidae, Galliformes) in Indonesia will serve to illustrate the point. Nesting grounds of the Moluccan megapode (*Eulipoa wallacei*) on Haruku island are still managed traditionally, and evidently sustainably, by local communities who lease the nesting grounds on an annual basis to the highest (resident) bidder. The fee is used for community requirements and there are rules to ensure the perpetuation of the nesting colonies. In contrast, nesting grounds of the maleo megapode (*Macrocephalon maleo*) in North Sulawesi are plundered unsustainably and some have been abandoned because local traditions have broken down in a province influenced and altered by the arrival of Javanese and Balinese transmigrants (Argeloo and Dekker 1996). In traditional use of wildlife, sustainability and the stability of communities are obviously closely linked.

Human population density, regardless of traditions, may be decisive in determining the sustainability of wildlife use at community level. If, when a cake is shared out fairly, there are only a few crumbs each, nobody is likely to be satisfied. In India it is not uncommon to find many thousands of people claiming traditional rights to the produce of a small remnant forest. Under these circumstances the conservation of biodiversity is unlikely to be served by simple schemes aimed at sharing out the instrumental benefits of the remaining wildlife. It could be critically important to ensure that any returns are spent on projects that have wide local support.

15.3.4 The institutional component

As Figure 3.5 indicates, this is the central component of the sustainability triangle insofar as it represents the mechanisms through which the other three components must interact for overall sustainability. The institutions will include administrative, organizational and financial capabilities. The word 'capabilities' needs emphasis because rules and regulations are only as effective as the way in which they are administered and enforced. In Tanzania, for example, it has been estimated that about 60% of wildlife utilization is illegal and much of the use of wildlife throughout Africa is technically against the law (Milner-Gulland and Leader-Williams 1992). In Kenya it has been claimed that illegal use of wildlife is at such a level that it threatens to cause catastrophic declines in wildlife populations. Kenyan game ranching schemes have had limited success because of a lack of marketing infrastructure and restrictions on the use of by-products (Kock 1995).

From the examples mentioned in this book (EFTA for the butterfly cottage industry; ICLARM-CAC for giant clams; CAMPFIRE for community use of African game; government agencies and sporting associations for field sports) it is clear that a variety of institutional arrangements can be applied successfully to the use of wildlife, and there are proponents of both private and public ownership – each with success stories to support their cases. There is consensus, however, on the unsustainability of using wildlife with the freedom of open-access commons. When nobody feels responsible for a resource, everybody feels free to take advantage of it. Any situation in which people are inclined to say that 'if we don't finish it off somebody else will' is extremely unsatisfactory for conservation.

Notwithstanding the successes under different ownership regimes, the issue of ownership is often at the root of institutional problems. In Australia, for more than a decade Gordon Grigg has been promoting the sustainable use of kangaroos for conservation purposes. But Australian conservation groups object to the principle of private citizens owning wildlife, while graziers, on the other hand, are concerned about becoming reliant on a resource that they cannot own. Grigg (1995) suggests that a possible resolution would be to develop a tagging system whereby landowners could be supplied with tags for their approved quota of kangaroos. The tags would authorize all necessary contractual arrangements for the culling of the animals in the specified area. Ownership of a kangaroo would not be legally transferred to a tagholder until a tag was attached to the carcase.

Even under private ownership the mobility of some species can present problems unless neighbouring landowners can cooperate. If properties can be amalgamated for wildlife management purposes, there will be less incentive for a landowner to kill an animal, whether it be a partridge or a moose, before it moves away to become the property of a neighbour.

15.4 INTERVENTIONS AND MANIPULATIONS

Examples could be found to show every level of intervention from broad-spectrum hunting and gathering in which the integrity of natural ecosystems is largely left intact, to domestication and selective breeding in environments so modified by man that it is difficult to see what the original landscape may have looked like. As a generalization, biodiversity is decreased along the progression indicated by Figure 15.1. Where destruction is a matter of degree, and a utilization scheme is helping to prevent more severe impacts from alternative forms of land use, the conservation advantages need to be assessed in the broad context of local and regional planning – or lack of planning as the case may be.

Figure 15.1 suggests only the range of consumptive use but the impacts of using wildlife non-consumptively can also be high and could conceivably be

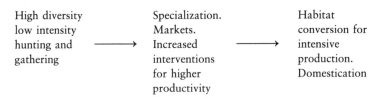

Figure 15.1 Decrease of biodiversity with increasing specialization and intensity of use.

more destructive than light consumptive use. A disturbance which causes birds to desert their nests, for instance, would be more serious for the birds than would a sustainable harvest of eggs or chicks from the same nests. The consumptive use of animals usually necessitates deliberate killing because it is not often possible to collect produce from completely wild animals in the way that we take wool from a sheep or milk from a cow. Exceptions include the collection of moulted 'wool' from musk oxen (*Ovibos moscha-tus*), feathers from the nests of various eider ducks and guano from some large bat and bird colonies. Very few extractive procedures have proved successful in wild animals. The collection of snake venom by the Irula Co-operative Venom Centre in India appears to be unique in that the snakes are released unharmed after about 3 weeks in captivity. During that time venom is extracted for antivenin production. Capture and handling are expertly done by Irula tribesmen (Whitaker and Andrews 1995).

It is possible to categorize utilization schemes according to the kinds of interventions and manipulations being applied, but it is doubtful whether the categories would be very useful in predicting conservation benefit or sustainability. This is because in practice so many other variables are involved. Table 15.2 sets out categories of intervention which include examples found in this book.

15.5 MONITORING

The basis of adaptive management is monitoring but that word is used so frequently and casually that there is a risk of its being treated as a token acknowledgement that we ought to be keeping an eye on things. In examining the requirements and some of the many techniques of monitoring, Goldsmith and his colleagues (Goldsmith 1991) make it clear that, to the extent that monitoring is a matter of keeping a watchful eye, it must be a scientific and purposeful eye.

Monitoring has been defined as 'intermittent (regular or irregular) surveillance carried out in order to ascertain the extent of compliance with a predetermined standard or the degree of deviation from an expected norm' (Hellawell 1991).

A monitoring programme therefore differs from ordinary surveillance or surveys in that it is not open-ended but is intended to produce findings in relation to preconceived expectations, however vague they might be. Accordingly, as the first point in his 19-point checklist for a monitoring programme, Goldsmith asks the question: 'Are your objectives clear?' His second point requires that the objectives be ranked in order of priority.

In connection with wildlife harvesting there will be a need to monitor population responses of the target species and, because the precautionary principle should have been followed, the expectation will be that the harvesting has no detrimental effect. There may be several other monitoring

Table 15.2 Categories of wildlife management interventions

Management intervention	Examples (varying degrees of sustainability)
Harvesting only. No habitat modification or protection. No interventions to increase productivity	Whaling. Saiga antelopes. Diversity of hunting/gathering
Selective, controlled harvesting for population management but no habitat manipulation or protection	Deer, other large mammal herds. Seals. Tegu lizards (Argentina). Gull colonies (eggs)
Habitat destruction resisted	Muttonbirds, megapodes. Deer, other large mammal herds. Fish, ducks for field sport
Active measures to protect/extend natural habitat	Deer, other large mammal herds. Partridge, grouse, ducks for field sports. Fish
Specific habitat modifications for increased productivity of target species	Butterfly ranching, wild bees (nest boxes). Oyster beds, mussel rafts. Game species (winter feed, range management, predator control)
Seeding the wild from captive bred stock	Inland fisheries, shellfish (experimental). Some game birds
Harvesting from the wild and growing to commercial stage in captivity (ranching as defined by IUCN)	Crocodiles. Turtles. Tuna fish (South Australia)
Harvesting introduced pests (with/without benefit to conservation of native wildlife)	Deer (New Zealand, Australia). Monkeys (Mauritius). Hares (Argentina). Goats, pigs, others (Australia)
Capture of wild animals for extraction of products; animals subsequently released	Very few examples. Snake venom (Irula project). Arrow frog poison (small South American communities)
Collection of non-living products from wild animals; no capture involved	Guano. Discarded shells. Eider-down, musk-ox 'wool'. Birds' nests (for soup)
Harvesting fenced (otherwise genetically separated) animals from wild populations	Reindeer. Southern African farmed game. Fish, bullfrogs
Breeding of captive animals, including new selective breeding/new domestications	Ostriches, emus. Iguanas, tegu lizards, crocodiles. Grasscutter, capybara. Fish, shellfish (mariculture)
Expertise gained from captive use employed to conservation advantage	Education value of public exhibits. Falconry, aviculture. Keeping aquaria/vivaria. Research, reintroductions
Wildlife viewing/photography (ecotourism)	Mainly conspicuous mammals and birds or animals of special interest

objectives to be ranked, such as increase of species richness in conservation headlands, or woodland regeneration in deer forests, or improved relationships between local residents and wildlife managers.

But how does one distinguish the effects of harvesting from other variables, such as seasonal fluctuations? Where, when and what should one sample? And how does one decide when to stop? These and many other questions must be answered before a monitoring programme can be implemented. At the same time the constraints of monitoring must be accepted. It may be useful to combine experimentation with monitoring; for example, by monitoring inside and outside animal exclosures. But there will always be limits and sometimes no monitoring programme could possibly have the statistical power to detect change within the time frame that managers are forced to accept. Under these circumstances it is still preferable for those involved to make decisions on the basis of informed guesswork, and such data as are available, than to have the decisions made by politicians for reasons that may have nothing to do with conservation.

15.5.1 Species and population change

Assuming that clear objectives have been set and a well planned monitoring programme has been implemented, how is management to respond to the results? Will there be specified criteria for taking certain management actions? Will it be possible to detect population changes before unacceptable damage has been done? To take an extreme example, it would be grossly irresponsible to harvest adult green turtles (*Chelonia mydas*) to the point where a monitoring programme began to show a population decline. By then, recruitment could have been disrupted and it could take several decades to restore the damage because green turtles do not breed until they are about 30 or more years old. Marine turtles are very difficult to monitor at sea, and because the proportion of breeding females fluctuates from year to year, there is also a poor correlation between numbers nesting and the numbers of potential breeders (Limpus and Nicholls 1988). In reality there is no harvestable surplus of adult green turtles and the species is already endangered by the pressure of subsistence hunters and the incidental losses caused by drownings in fish nets and shark barriers (Marsh 1995).

There are some circumstances, such as those described for Scotland's red deer (Reynolds and Staines, Chapter 10, this volume) in which a reduction in population size may be the management objective. The degree of acceptable change can then be specified. As it happens, red deer are also good subjects for monitoring.

Marsh (1995) lists the characteristics of animal populations which make for satisfactory monitoring and the detection of change:

• having a lifespan of years rather than decades or months

- having a life history which is well understood – especially the relationship between fecundity and population size
- occurring at population levels which allow changes in abundance to be detected affordably and with high power
- being relatively sedentary and always available to be counted
- being amenable to experimental manipulation so that changes in abundance can be investigated.

15.6   FINALLY

The total human impact upon wildlife cannot be diminished in the foreseeable future. While the bulk of the destruction can be regarded as the collateral damage of human development, much of the deliberate killing is also beyond or outside the control of official paperwork. Whereas an endangered species can only move between the zoos of two nations if it is accompanied by a thick sheaf of forms and permits, the same species in the wild could well be hanging from a stick on its way to the meat market.

The attempt to derive conservation advantage from using wildlife should not be seen as a bold new conservation initiative for it is not proactive at all; it is a concerned response to an existing, and worsening, situation. Nothing in this book is intended to suggest that the full value of wildlife can be measured in terms of money or any other unit. The various economic valuations (consumer surplus, option value, existence value, bequest value) all have their uses but none can reveal the total worth of wild animals. Funtowicz and Ravetz (1994) expressed the point perfectly when they wrote: 'The worth of a songbird definitely has its monetary aspect but the endangered songbird is not thereby reduced to a commodity . . .'.

Neither does this book propose that the controlled use of wildlife is a panacea for conservation; on the contrary, a great many difficulties have been emphasized. And yet there are examples of wildlife being used sustainably, and with conservation benefits, so we know that utilization can sometimes help. There will be many more possibilities in all parts of the world. By asking the right questions we can hope to identify them. By proceeding cautiously we can hope to avoid ecological mistakes. The structure of this final chapter suggests the sort of questions that must be asked. Table 15.3 puts some of them in general terms; more precise questions can be asked in relation to real proposals in actual locations.

It is sometimes remarked that using wildlife is not conservation but this only amounts to expressing an opinion as to how conservation should be defined. The word conservation is not like a technical term from physics or mathematics; it has no precise definition. But conservationists often urge us to try to tread more lightly upon this planet so perhaps we should use that metaphor. We may then think of conservation in terms of our collective

Table 15.3 Planning components of a wildlife utilization scheme

Component	Sub-components	Some general questions
Alternatives to scheme	Land/wildlife Certainty/uncertainty Public opinion	What other threats to land and/or animals? How probable are the other threats? Local opinions on proposed alternatives?
Conservation benefits of scheme	Direct/indirect Wildlife numbers Wildlife security	Will revenue be raised/used for conservation? Predicted effect on target species/ others? Will legal status of habitat/wildlife be improved? Will local attitudes to wildlife improve? Risks relative to alternatives?
Ecology	Niche/life history Population growth Behaviour	Possible effects on other species? Intrinsic rate of increase? Can most expendable stage be targeted? Density dependence? Spacing behaviour? Seasonal movements? Appropriate scale of operations for ecological sustainability?
Interventions	Needs Opportunities	Essential/desirable/possible management needs of populations and/or habitat?
Status of species	Scarcity/abundance pest/ resource	How is target species regarded by locals? Are different management goals compatible?
Institutional component	Resource tenure Administration Organization Finance	Local ownership/secure tenure? Adequacy of existing structures? Capabilities, training needs? Financial resources? Accountability? Honesty?
Socio-economic component	Markets Profitability Incentives	Local/other markets established? Assessed? Adequacy of rewards relative to social/economic status of those involved?
Community component	Lifestyle/traditions Numbers involved	Dependable or transitional lifestyles? Can resource be acceptably appropriated or effectively shared?
Monitoring	Objectives Plan Resources	Are objectives agreed? Is plan agreed and does it meet objectives? Are funds/resources assured for plan duration?

contortions as we all try to tread more lightly while at the same time half of us are struggling to acquire substantial boots and every year another 90 million people join the march. Thought of in this way, conservation, and conservationists, are in no position to be philosophically narrow. Shackles, even metaphorical ones, are the last thing we need.

REFERENCES

Allen, L., Hynes, R. and Thompson, J. (1995) Is commercial harvesting compatible with effective control of pest animals and is this use sustainable?, in *Conservation through Sustainable Use of Wildlife* (eds G.C. Grigg, P.T. Hale and D. Lunney). Centre for Conservation Biology, The University of Queensland, pp. 259–66.

Argeloo, M. and Dekker, R.W.R.J. (1996) Exploitation of megapode eggs in Indonesia: the role of traditional methods in the conservation of megapodes. *Oryx*, 30 (1), 59–64.

Cairns, S.C. and Grigg, G.C. (1993) Population dynamics of red kangaroos (*Macropus rufus*) in relation to rainfall in the south Australian pastoral zone. *Journal of Applied Ecology*, 30, 444–58.

Caughley, G. (1993) Elephants and economics, *Conservation Biology*, 7 (4), 943–5.

Caughley, G. and Sinclair, A.R.E. (1994) *Wildlife Ecology and Management*. Blackwell, Cambridge, Mass.

Choquenot, D., O'Brien, P. and Hone, J. (1995) Commercial use of pests: can it contribute to conservation objectives?, in *Conservation through Sustainable Use of Wildlife* (eds G.C. Grigg, P.T. Hale and D. Lunney). Centre for Conservation Biology, The University of Queensland, pp. 251–8.

Elliott, T.K. and Woodford, K. (1995) Ecology, commerce and feral goats, in *Conservation through Sustainable Use of Wildlife* (eds G.C. Grigg, P.T. Hale and D. Lunney). Centre for Conservation Biology, The University of Queensland, pp. 267–75.

Eltringham, S.K. (1994) Can wildlife pay its way? *Oryx*, 28 (3), 163–8.

Funtowicz, S.O. and Ravetz, J.R. (1994) The worth of a songbird: ecological economics as a post-normal science. *Ecological Economics*, 10, 197–207.

Goldsmith, F.B. (1991) Synthesis, in *Monitoring for Conservation and Ecology* (ed. F.B. Goldsmith). Chapman & Hall, London, pp. 269–71.

Grigg, G. (1995) Kangaroo harvesting for conservation of rangelands, kangaroos . . . and graziers, in *Conservation through Sustainable Use of Wildlife* (eds G.C. Grigg, P.T. Hale and D. Lunney). Centre for Conservation Biology, The University of Queensland, pp. 161–5.

Hellawell, J.M. (1991) Development of a rationale for monitoring, in *Monitoring for Conservation and Ecology* (ed. F.B. Goldsmith). Chapman & Hall, London, pp. 1–14.

Jori, F., Mensah, G.A. and Adjanohoun, E. (1995) Grasscutter production: an example of rational exploitation of wildlife. *Biodiversity and Conservation*, 4 (3), 257–65.

Kock, R.A. (1995) Wildlife utilization: use it or lose it – a Kenyan perspective. *Biodiversity and Conservation*, 4, 241–56.

Limpus, C. and Nicholls, N. (1988) The southern oscillation regulates the annual number of green turtles (*Chelonia mydas*) breeding around northern Australia. *Australian Wildlife Research*, **15**, 157–61.

Luxmore, R. and Swanson, T.M. (1992) Wildlife and wildland utilization and conservation, in *Economics for the Wilds* (eds T.M. Swanson and E.B. Barbier). Earthscan, London, pp. 170–9.

Marsh, H. (1995) The limits of detectable change, in *Conservation through Sustainable Use of Wildlife* (eds G.C. Grigg, P.T. Hale and D. Lunney). Centre for Conservation Biology, The University of Queensland, pp. 122–30.

Milner-Gulland, E.J. and Leader-Williams, N. (1992) Illegal exploitation of wildlife, in *Economics for the Wilds* (eds T.M. Swanson and E.B. Barbier). Earthscan, London, pp. 195–213.

Mrosovsky, N. (1989) Natural mortality in sea turtles: obstacles or opportunity? NOAA Technical Memorandum NMFS-SEFC-226. Proceedings of the Second Western Atlantic Turtle Symposium (ed. L. Ogren), pp. 251–64.

Ojasti, J. (1991) Human exploitation of capybara, in *Neotropical Wildlife Use and Conservation* (eds J.G. Robinson and K.H. Redford). University of Chicago Press, Chicago, pp. 236–52.

Spinage, C.A. and Matlhare, J.M. (1992) Is the Kalahari cornucopia fact or fiction? A predictive model. *Journal of Applied Ecology*, **29**, 605–10.

Thomsen, J.B. and Brautigam, A. (1991) Sustainable use of neotropical parrots, in *Neotropical Wildlife Use and Conservation* (eds J.G. Robinson and K.H. Redford). University of Chicago Press, Chicago, pp. 359–79.

Webb. G.J.W. (1995) The links between wildlife conservation and sustainable use, in *Conservation through Sustainable Use of Wildlife* (eds G.C. Grigg, P.T. Hale and D. Lunney). Centre for Conservation Biology, The University of Queensland, pp. 15–20.

Whitaker, R. and Andrews, H.V. (1995) The Irula Co-operative Venom Centre, India. *Oryx*, **29** (2), 129–35.

Index

Page numbers appearing in **bold** refer to figures and page numbers appearing in *italic* refer to tables.